Australia's Dangerous BUSH CREATURES

Myke Mollard

For Benjamin

Red-bellied Black Snake

In my adventures I've come across these snakes many times. Finding them to be gentle, placid creatures, always sunning themselves in familiar patches of bush. Hardly dangerous? But I'd never push the friendship.

Non-aggressive, the **Red-bellied Black Snake** will usually slither away before striking. Native to Australia, this venomous velverty-black snake has no recorded deaths to its name. It prefers feasting on frogs and is known to eat other young snakes.

Australia's Dangerous Bush Creatures

Australia is home to some of the most dangerous animals in the world. We have the most venomous snakes, spiders, fish, jellyfish, a deadly octopus and many sea snakes. Great White Sharks patrol our coastlines, Saltwater Crocodiles inhabit our rivers and both can definitely tear you limb from limb. Yet as fearsome as this all sounds, you will soon realise that all these iconic Australian bush creatures are both fascinating and complex. This hand-picked selection of animals illustrates the spectacular biodiversity of our land as well as its dangers. I hope that this gallery of 'wildlife with deadly reputations', infused with accounts of my encounters with them, will excite your imagination and help you explore how wild and fascinating nature truly is.

A Trio of Highly Venomous Copperheads

When I was thirteen, most weekends I hiked through the High Country, up mountains like Mt. Cobber, Mt. Howitt, The Devil's Staircase, The Razor, The Viking, Mt. Buggery, The Bluff and Eagle's Peak. In the cracks and grassy crags we often came across coiled Alpine Copperheads, also known as Highland Copperheads. We'd catch them slithering through the sun-bleached grass, the dry crackle of dead leaves and curled gum bark underfoot seeming not to scare them. Snakes don't hear you coming, but they feel vibrations and with forked tongues they can surely smell our sweaty passage. Climbing up the rocky tracks, I've come eye-to-eye with these highly venomous sentinels, but every time they've just keenly watched me, tongues licking the air, and I've calmly left them alone.

When you encounter snakes in the wild, it's best to stay calm and let them move along or stay out of their way. It doesn't take much to step around them. Snakes are a great sign of a healthy environment and in the High Country an indication you are close to water. Up on the ridges of the Great Alpine Trail, far above the rivers which cut deep in the valleys below, you have to find soaks and springs to fill your water bottle. They may be marked on a watershed map, but if there is water, there are frogs and with frogs you'll generally find snakes close by.

Mountain kings, **Copperheads** enjoy the colder climate and have the shortest hibernation of all our land snakes. **Alpine Copperheads** in particular, have been known to bask in sun even after snowfalls and again in the early spring melt.

All **Copperheads** are hardy, solid looking snakes. They vary in length and size depending on their location. There are three varieties: the **Alpine** or **Highland Copperhead** grows to around 1m; the **Southern** or **Lowland Copperhead** of Victoria is the largest at nearly 2m (hence why they are the most feared); and the **Pygmy Copperhead** of Kangaroo Island and the Fleurieu Peninsula, which only grows to 60cm.

Highly venomous, but not intentional or effective man-killers, **Copperheads** are often clumsy when displaying their defensive behaviour. With shorter fangs than the **Coastal Taipan**, **Copperheads** will not actually deliver a full 'fatal' dose of venom. Hopefully, this will dispel some myths surrounding these snakes, although all **Copperhead** snake bites should be taken seriously.

Human encounters can be common around farms, water sources and swampy grasslands, because these snakes hunt by day for frogs and small rodents. It's best to always respect them, be mindful when bushwalking and just calmly keep out of their way if you cross paths with these bush creatures.

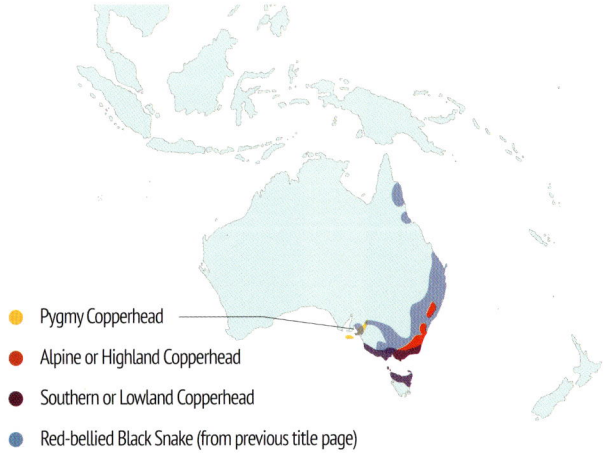

- Pygmy Copperhead
- Alpine or Highland Copperhead
- Southern or Lowland Copperhead
- Red-bellied Black Snake (from previous title page)

THIS PAGE
An Alpine Copperhead basks in the afternoon sun enjoying a brisk mountain breeze. It is coiled on dry leaf-litter, in a shallow sea of bracken and sun-kissed native grass.

Our Notorious Eastern Brown Snake

The Eastern Brown Snake is known as our biggest killer. It might not be the most venomous snake, but due to its numbers, range and close proximity to farms and urban habitation, these snakes are constantly at our doorstep. It's estimated that around 40% of all snakebite deaths in Australia are from this Brown Snake. With what is considered the quickest killing venom in the world, if left untreated their bite can kill in 30 minutes. That's pretty dangerous! No other snake in the world has killed so many people, so quickly and on such a regular basis. This Illustrates the importance of antivenom and being cautious in the Australian bush.

In my youth, I often visited farms across Victoria and New South Wales and my parents used to own a 'hobby' farm on the Mornington Peninsula. In those days, working on farms throughout the summer months, 'hay-carting' and 'jackarooing', I encountered many **Eastern Brown Snakes** both big and small. They always strike initial fear into horses and their presence can attract the attention of working dogs. Personally, I've tried to avoid these snakes, knowing the potential risk they pose.

The adult **Eastern Brown Snake** has a slender build and can grow to an impressive 2m in length. Found in most habitats, except dense forests, it prefers grassy farmland and the outskirts of urban areas because of the usual large populations of mice. A solitary hunter, active on hot days in the late afternoon, its behaviour often puts it in close contact with human activity.

As an active predator, it helps control overabundant rodent populations, but its diurnal activity makes it easy prey in turn for kookaburras, eagles and feral cats: these snakes are an important part of our diverse ecosystem.

Eastern Brown Snakes will normally avoid people, but if they are cornered, threatened, disturbed or surprised, they will strike out and defend themselves. If provoked or taunted, they can be extremely aggressive and, with a highly venomous payload, aren't afraid to teach humans a quick lesson in toxicology. Make no mistake, these snakes should be regarded as one of our most dangerous creatures.

On a world scale, the **Eastern Brown Snake** is considered the second-most venomous snake, behind only the **Inland Taipan** of central eastern Australia. The venom causes anti-clotting, hypotension, nausea and collapse. Symptoms can be rapid and severe, sometimes beginning in less than two minutes. Vomiting, sweating, abdominal pain, headache, acute kidney disfunction, possible seizures, internal bleeding and, within 20-30 minutes, a massive cardiac arrest is the final fatal blow. Scary! But these snakes will avoid you, so it's best to avoid them.

Appearance *The colour of its surface scales range from pale brown to black, while its underside is pale cream-yellow, with orange or grey splotches. It has a slender to average build and a rounded snout.*
Size *Most grow to 1.5m with some large individuals reaching 2m.*

● Eastern Brown Snake

THIS PAGE
A startled Eastern Brown Snake rears up, displaying that classic 'S' striking pose, a warning that it's not happy. Nearby, a Wedge-tailed Eagle has spotted its basking spot amongst the dry grass.

Australia's Dangerous Dugite Snake

Within this book I've only selected a handful of our most dangerous snakes. Australia has around 140 species of snakes and some 100 of these are considered venomous, although only a dozen could kill you. Worldwide there are about 25 venomous snakes that pose a real threat to human life. What freaks out many visitors to Australia is that half of these snakes can be found here. In the arid, sandy hinterland hugging the south-western Australian coastline, basks the dangerous Dugite Snake or Spotted Brown Snake.

In the wild, **Dugites** shelter beneath logs and rocks. When disturbed, they often slither away, but can become highly agitated, excitable and aggressive if provoked: they are known to defend themselves if cornered. These snakes are active in the daylight hours of dawn, mid-morning and sometimes at dusk. On hotter days, they will find a nice piece of iron or log-hollow to rest in. More active in their breeding season around October/November, the **Spotted Brown Snakes** are like most **Brown Snakes**, roaming where rats and mice, their favourite prey, are plentiful. This brings them close to human habitation in both urban and semi-rural areas, around houses, backyards, sheds and farm buildings.

As a species of **Brown Snake**, the **Dugite's** venom is one of the most lethal in the world, causing similar effects to the previously mentioned **Eastern Brown Snake**. This venom was never intended for humans, being perfectly evolved to quickly dispatch and disable small mammals and rodents: **Dugites** will try to avoid biting larger mammals like us.

Urban expansion, in and around Perth in the last two decades, has brought our paths closer together. Now the **Dugite** has formed a infamous reputation, making local headlines in Fremantle in 2015 with one death and many more snake bite victims which recovered from the experience. Locally, approximately 70% of all snake bites reported to Perth hospitals are attibuted to the **Dugite Snake**. Thanks to doctors prompt use of antivenom, fatalities like Fremantle's are rare.

Because of the snake's size and highly toxic venom, the **Dugite** is considered to be very dangerous, deserving its place on Australia's most dangerous bush creatures list. Fortunately for humans, the fangs are actually quite small and just wearing solid shoes, heavy denim fabrics and other appropriate clothing is import if out bush in south-western Australia. Regardless of these smaller fangs, anyone with a suspected bite should seek immediate medical attention.

Appearance *Dugite colours vary widely between individuals, so it can be an unreliable means of identification. The body is long and slender in build. The head is rather small and indistinct from the neck and has a strong brow ridge over each large eye, which are blackish brown with a golden orange rim surrounding a round pupil. The mouth lining is pink, while the tongue sheath and throat are black. There are three subspecies, all fairly similar in general appearance. Mainland subspecies' overall colour ranges from brown, through olive brown to brownish grey, with irregular black/dark grey spots. Eastern and southern individuals are heavily spotted. Scales are relatively large with a semi-glossy appearance, generally pale grey to brown with brownish orange or dark grey blotches. There are also two different island subspecies, isolated from the mainland, which are generally smaller and have a more uniform blackish brown appearance above their lighter coloured underbellies.* **Size** *A Dugite's body is long, growing up to 2m in total length including tail, but the typical size is around 1.5m.*

● Dugite or Spotted Brown Snake

OPPOSITE PAGE
Although naturally shy, the Dugite is easily agitated when disturbed and will raise its forebody upright in a tight S-shape before striking.

Our Venomous Coastal and Inland Taipans

The Inland Taipan holds the reputation of being one of the world's most venomous snakes. It was once rarely encountered, due to its habitat locations, and Australia didn't have an antivenom for a long time, until the 1950s. Then a snake collector spotted a Coastal Taipan at the rubbish tip in suburban Cairns. He caught and bagged it, but in the process was bitten several times. Rushed to hospital he was given Tiger Snake antivenom, but the man had no chance and died. While he was still able to speak, he insisted that the killer Taipan be sent to Melbourne to the serum laboratories for milking. Research began and within five years the first life was saved by his dying wish.

Before the antivenom was developed, **Australia's Taipans** were the most feared snakes in Australia: a bite from these snakes would mean certain death for anyone, even a fully grown man. The only known survivor before 1950 was a bitten through a thick leather boot and socks! The **Taipan's** bite, although having the longest fang length of any Australian snake, didn't fully penetrate the skin and, with some quick thinking, his colleagues promptly bled the wound and he luckily survived.

First Australians have said that to survive a **Taipan** bite you must put yourself in a meditative coma, a trance-like state, lying stiller than still for three days until the venom passes through you. Both bleeding the wound and inducing a trance are no longer the recommended first aid methods, but not panicking, staying as immobile as possible and proper first aid techniques can keep you alive until you receive treatment and the antivenom.

There are two types of **Taipans** in Australia and both are equally venomous and highly dangerous: the **Inland Taipan** and **Coastal Taipan**. They should be left alone and treated with the upmost respect at all times. They have Australia's fastest striking speed, 12mm fangs (one of the longest of our Aussie snakes) and they deliver a venom load four times that of a **Tiger Snake**. If that knowledge isn't a warning to stay back, then maybe their S-shaped striking pose (when they feel threatened) should give you a good indication of their well-deserved global reputation.

It is beautiful, sleek, agile, totally deadly, often unpredictable and aggressive, but definitely a snake ready to defend itself if threatened or challenged.

You will find the **Inland Taipan** in the stoney and semi-arid regions of central eastern Australia. The main diet of these **Taipans** is rats, small mammals and species of mice. In the modern world, this does bring them into urban areas, industrial areas and farms seeking a easy feed.

Be mindful that when bushwalking, holidaying, camping, traveling through or working in well-grassed tropical woodlands, the cane fields and farms on the Queensland coast, you are in **Coastal Taipan** territory. **Coastal Taipans** are also found on the Northern Territory coast.

Appearance *The Taipan has a very distinctive, angular shaped head, long and narrow, with orange-red eyes slightly inset under the edgy browline. Their long slim body is coloured uniformly either brown, tan or fawn. The Inland Taipan is often a darker russet brown to nearly black in some locations. The flat cap of their head is coloured like their body, but the snout and face are cream, lightest on the nose. This colour extends to their belly, cream to yellowish with sometimes reddish marks near the throat.*
Size *It is said that these snakes can reach over 3m in length, but their average size is around 2.5m, which is still a pretty large venomous snake.*

- Inland Taipan
- Coastal Taipan

THIS SPREAD
Poised to strike! The much feared Coastal Taipan.

OPPOSITE PAGE
Lightly-coloured and subtly banded Northern Death Adder.

THIS PAGE
Up close and deadly! A beautiful Common Death Adder.

Death Adder's, deadly art of Camouflague

Part of my desire to draw wildlife is to help kids to understand and identify our bush creatures. A parent messaged me the other day with an amazing story: their six-year-old saw a Death Adder while on holiday and warned her father and other tourists about it, saving the day. It was highly camouflaged, but spotting this deadly and highly venomous snake created a great outcome for both the onlookers and the Death Adder - one of safety and mutual respect.

The **Death Adder** is a master of camouflage due to its colour and distinctive banded scales. It stays perfectly still, hiding in loose leaf litter or woodland debris. It prefers warm grassy scrub land and has a worldwide reputation as one of the most venomous snakes.

Unlike most snakes, the **Death Adder** is a pure ambush predator. Coiled up, it makes itself quite inconspicuous then lies in wait for its prey, some times for days. Like the Pit-Viper it twitches its grub-like tail to lure in its prey.

A non-aggressive snake, the **Death Adder** is really dangerous because of its hunting style and beautiful camouflage. The fact is, most humans often don't even notice this deadly adder until it's too late.

Possessing the longest fangs of any Australian snake, the **Death Adder** is a snake to watch out for in the bush. Fun fact: despite looking like a **European Adder**, the **Death Adder** is actually a member of the Elapidae snake family, not the Viperidae family, which are not found in Australia. There are eight types of **Death Adders**: the **Common Death Adder**; the **Southern** and **Northern** varieties; the **Barkly Tablelands**, **Kimberley** and **Desert Death Adders**; plus the **Smooth-scaled** and **Rough-scaled Death Adders**.

Appearance *The Death Adder has a broad, flat triangular head and a thick body, with bands of red, brown and black, a grey, cream or pink belly and tapered worm-like tail.*
Size *It can reach a body length of up to 1m, but the average length is around 65cm.*

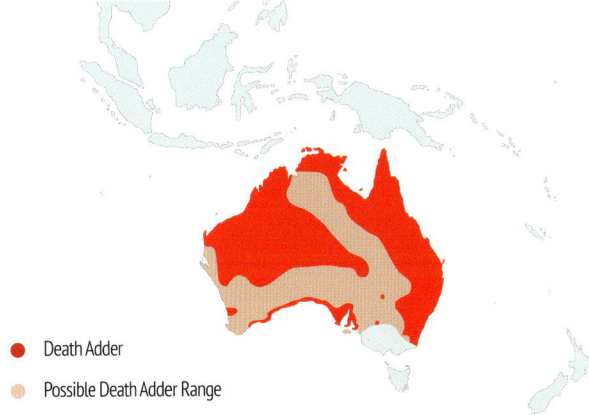

- Death Adder
- Possible Death Adder Range

Our Fearsome Tiger Snakes

Growing up in Melbourne and regularly visiting farms in regional Victoria, I was always warned how dangerous Tiger Snakes were. As a kid, I was often told to be careful; when walking through long dry grass, playing down near creeks and streams, 'cubby-housing' in haysheds, lifting up old tin sheets, playing on idle farm equipment. In summer, these were perfect places to encounter a Tiger Snake.

The Australian **Tiger Snake** has a fearsome reputation. Highly venomous, aggressive by nature and not afraid of humans, these fast, agile snakes, have given rise to some great stories. Like the **King Cobra**, they flatten their heads and necks as a warning that they are going to attack or strike, and should be avoided at all times.

What makes **Tiger Snakes** quite deadly is that they are quite used to us, often living right under our noses or feet. They are quite active on a summer's day especially at dusk. Preferring moist wetlands and swampy areas with good rainfall, frogs are a **Tiger Snake's** favourite food, although, mice, small lizards, and other young snakes are preyed upon too. They are good swimmers and can cross rivers and streams. They are also great climbers, hunting fairy wrens and raiding birds' nests for their hatchlings and eggs.

There are only two main colour types of **Tiger Snake**: the **Common Mainland Tiger Snake**, which is brown in colour, and the **Black Tiger Snake** of Western Australia, and the Yorke and Eyre Peninsulas. **Black Tiger Snakes** also live on King Island, Kangaroo Island and Tasmania. A nomadic snake, they move about looking for mates. Prolific breeders, all **Tiger Snakes** birth up to 20-30 live young per litter which they abandon shortly after birth to fend for themselves.

Appearance *This is quite a heavily built snake, with around 40-50 distinctive yellow or cream tiger-like stripes down its body. It has a broad head, with a wide shield-like head plate between its eyes. The overall colour of the Tiger Snake is either brown, green/brown or black. It has a cream, yellow or olive-grey belly and black or dark brown eyes.* **Size** *Some grow to a length of nearly 2m, but on average a Tiger Snake grows to around 1m.*

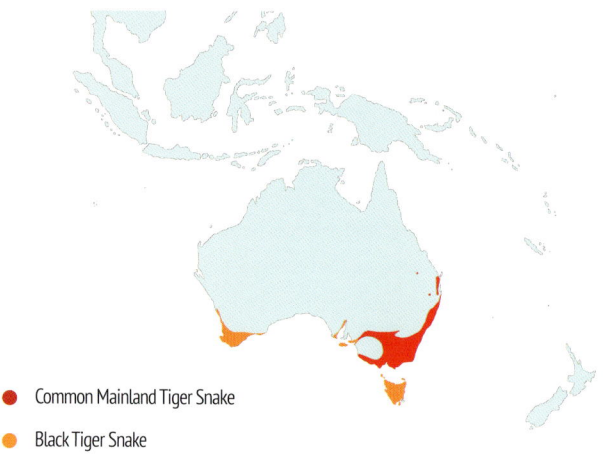

- Common Mainland Tiger Snake
- Black Tiger Snake

OPPOSITE PAGE
Western Australian Black Tiger Snake.

THIS PAGE
Common Mainland or Eastern Tiger Snake.

THIS SPREAD
On a cool overcast day, a large 3 metre King Brown Snake is on the move.

The Enormous Mulga Snake or King Brown

The mighty King Brown is Australia's largest venomous snake, reaching over three metres in length. That's huge! Endemic to northern, western and central Australia, it stays close to water sources, hidden springs and moist areas scattered through our sandy deserts and mulga bushlands. Despite the common name and coppery appearance, suggesting they are Brown Snakes, they're actually more closely related to our Black Snakes. Today it's more commonly known as the Mulga Snake. It sports a diversity of colourations, ranging from russet, coppery tan, reddish-green and brown to dark brown or even a dark blackish charcoal.

Mulga Snakes, traditionally called **King Brown Snakes**, are dry habitat generalists. They are comfortable in open woodlands, scrub filled with hummock grass, tightly packed stony 'gibber' desert or the shifting dunes of sandy deserts nearly devoid of vegetation. Within these harsh, sunburnt landscapes the **Mulga Snake** has a knack of finding soaks, springs and areas of greater moisture. The snake has a strong cultural significance to our first Australians. Indigenous trackers, when hunting, choosing camps or on walkabout, would have used these snakes' movements to follow ancient watercourses and tap into the Great Artesian Basin.

To conserve its own moisture, the **Mulga Snake** will do most of its hunting between dusk and late evening. During the middle of the day and between midnight and dawn, it retires to crevices in the soil, old animal burrows or under rocks or logs. During the warmer months its activity extends later into the evening and, when in cooler climates, it will venture out in the middle of the day.

The **Mulga Snake** is considered a highly venomous Elapidae species and a very dangerous snake if stumbled upon.

While its venom is not as potent as **Taipans** or **Eastern Browns** (think more **Black Snakes** and **Tiger Snakes**), if delivered in large enough quantities, without proper treatment, one bite can still have severe effects. Their venom mainly fatigues muscle tissue, causing paralysis from muscle damage with extensive pain and swelling at the bite site. In severe cases, without **Black Snake** (not **Brown Snake**) antivenom, a **King Brown Snake** bite will kill you.

In the past, numerous deaths from **Mulgas** have been recorded, especially before the availability of antivenom. Even with antivenom in major hospitals, getting bitten in a remote location, with only first aid and no way to get medical help fast, could still prove fatal. Although, Australia's most recent death from a **Mulga Snake** was over fifty years ago.

The **Mulga Snake** may not be the the most venomous or the most dangerous. It's name **King Brown** may trick you into thinking it's a 'brown' when it's a 'black snake'. But in the wild, it will always be our largest, most impressive looking venomous snake.

All Brown Snakes have a highly venomous bite, but not all Brown Snakes look alike.

Gwardar or Western Brown Snake

What a great name - **Gwardar!** It means 'go the long way around' in a First Australian's language: avoidance is key in this case. Sleek, agile and with a nasty reputation for being quite aggressive when disturbed or threatened, the **Gwardar**, or **Western Brown Snake**, is highly venomous and totally dangerous. Thankfully, like the lethal **Inland Taipan** and the elusive **Fierce Snake**, it lives in very remote areas, scattered across Western Australia.

Identification is complicated as it comes in a wide variety of colours, patterns and markings. Individuals from northern regions have tan upper parts, while those in southern areas are dark brown to blackish. Some have black banding, others are paler with speckles, some have black heads or alternatively ones that are tan and dark brown with spots or dots.

The **Western Brown Snake** is diurnal (active during the day) in winter, enjoying sunlight hours to hunt, and nocturnal during the warmer months. They prey upon small mammals and reptiles, including lizards and mice. Northern varieties are known to kill their prey with a combination of venom and constriction and, although this is not common, they can also be cannibalistic, eating their young and other **Brown Snakes**. A bite from any species of **Brown Snake** should be treated as life-threatening and medical attention sought without delay.

Appearance *It has various colourations, from very plain to shades of orange-brown with flecks and bands. Its belly is cream to orange with pink blotches. Some individuals have jet black heads.* **Size** *It generally grows to 1.8-2.5m in length, with the largest reaching up to 3m.*

● King Brown Snake or Mulga Snake
● Gwardar or Western Brown Snake (Including Northern Western Brown Snake)

OPPOSITE PAGE
A Mulga Snake or King Brown Snake posed to strike and defend himself.

THIS PAGE
An intricately patterned, startled Gwardar or Western Brown Snake.

Our Sea Snakes and the Banded Sea Krait

There are around 70 known species of Sea Snakes found in our oceans and Australia is home to 32 of them. These true sea snakes, highly evolved to sea living, are important to our reef ecosystems. Highly venomous, most Sea Snakes try to avoid all human contact, so bites are rare, although a fisherman did die recently while on a trawler off the Northern Territory coast. This was an unlucky accident as the Sea Snake was tangled in the net he was working on. This incident is believed to be the first Sea Snake fatality in more than 80 years.

The **Yellow-bellied Sea Snake** is a large venomous species and one of the most widely distributed snakes in the world. Boldly coloured, completely pelagic (living in the open sea), it uses surface currents, sea drifts and storms to get around the world's oceans. **Yellow-bellied Sea Snakes** rarely come to land, using floating mats of sea kelp for hunting, mating and protection. But, for some reason, **Yellow-bellied Sea Snakes** frequently get beached on Australian shores, washed up exhausted or sometimes injured. Christmas Island and the Cocos (Keeling) Islands, within Australian maritime waters, also have numerous beachings and healthy populations in their off-shore reef systems. These unmistakeable yellow and black **Sea Snakes** are easily spotted when beached. It's wise not to handle these delicate beached snakes as they are highly venomous and they can still defend themselves.

At least 22 species of Sea Snake have been discovered and identified in Western Australian waters alone. Just in Shark Bay the most common are the **Olive-headed Sea Snake**, or **Olive Sea Snake**, the critically endangered **Elegant Sea Snake**, also known as the **Short-nosed Sea Snake**, and the **Shark Bay Sea Snake**, which is unique to the area.

The **Olive-headed Sea Snake** is one of the larger aquatic snakes to call Shark Bay home and has a potent, neurotoxic venom for immobilizing prey. Catching fish is no easy task, so being highly venomous helps dispatch their prey in seconds. The **Olive-headed Sea Snake** actively hunts small to medium sized fish and invertebrates, including prawns and reef crabs.

The **Elegant Sea Snake** is slightly smaller, a medium-sized and slender snake growing to a length of 2 metres. Like most **Sea Snakes** it has a laterally compressed tail that aids in swimming. All these three snakes are very similar in size and build, but at the same time noticable features highlight their differences and diversity. **The Shark Bay Sea Snake** has a slightly different shape with narrow banding, the **Elegant Sea Snake** has a short snout and the robust **Olive Sea Snake** has a plain colouration with a distinctive olive head. Effects from a bite from any of these Elapidae **Sea Snakes** may include breakdown of muscle tissue, which can result in infections. Although highly venomous, **Sea Snakes** rarely attack divers or swimmers: shy and skittish, they would rather swim away from humans.

Australian reefs, including the Great Barrier Reef on the east coast and Ningaloo Reef on the west coast, are home to nearly half the world's known species of **Sea Snakes**. Sharing these reefs are **Sea Kraits**, which spend some of their time on land, the most distinctive being the black and white **Banded Sea Krait**. Not all **Sea Snakes** have the strength of venom to be dangerous to humans, but the **Banded Sea Krait** has a potent venom, hollow fangs at the front of the mouth and can deliver a nasty bite if cornered or feels it must defend itself. They prefer to hunt prey in nooks and crevices dotted throughout the reef. Once located, they envenom their prey and swallow it whole: digestion may take a few weeks. While **Sea Snakes** seem to hunt fish and crabs, the **Banded Sea Kraits** enjoy hunting and dinning on eels, specifically conger and moray.

Both **Sea Kraits** and **Sea Snakes** are admired by divers, snorkelers, marine biologists and oceanic photographers. It's safe to say that these shy and timid creatures don't have any intentions to bite or eat us. Their venom may be highly toxic and fatal, but most deaths and attacks are accidental and caused when people try to handle, save, corner one for a selfie or play with these aquatic wonders in the wild.

Honestly, that's just it - wild creatures are best left to their own devices. Humans have to learn to tread carefully through nature, be aware and respect the personal space of these dangerous bush creatures.

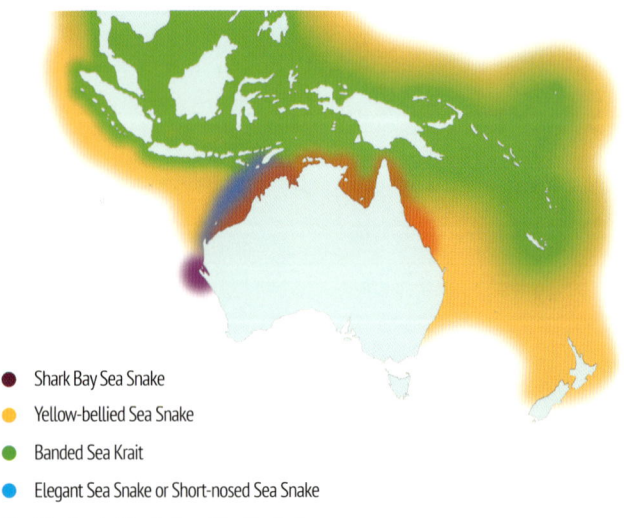

- Shark Bay Sea Snake
- Yellow-bellied Sea Snake
- Banded Sea Krait
- Elegant Sea Snake or Short-nosed Sea Snake
- Olive-headed Sea Snake or Olive Sea Snake

THIS PAGE
A strikingly adorned Yellow-bellied Sea Snake is in the centre. Around him, clockwise from the top, are: the narrow, cream-banded Shark Bay Sea Snake; two Elegant or Short-nosed Sea Snakes; a Black and White Sea Krait or Banded Sea Krait, slinking through the coral; two Olive Sea Snakes or Olive-headed Sea Snakes.

The Saltwater Crocodile

Out of all the dangerous bush creatures in this book, the Saltwater Crocodile is the only creature which regards people as prey... I just want to let that sink in!

Saltwater Crocodiles have thrived in Australia's lands and waterways for millions of years. They should be feared, and for good reason: there's a long history of them attacking humans who unwittingly enter their territory. With intimidating size and cunning camouflage they are expert ambush predators and powerful, primordial killers. Over short distances they can move with great speed on land and in the water they are agile swimmers and can launch themselves with explosive speed when attacking. Hunting their prey on mudflats, by the waters edge, and from beneath the water in both brackish rivers and clear saltwater, they are deadly. Should you have the unfortunate experience of becoming their prey, surviving an attack is pretty unlikely.

The **Saltwater Crocodile** is the largest living reptile known to science. Their remote ancestors diverged from other reptiles and started evolving into what they are today 240 million years ago. Fossil remains of the existing living **Saltwater Crocodile** species have been excavated in northern Queensland and dated to the Pliocene (5.4-2.4 million years ago): our wild crocodilian populations were here well before humans came along and started turning these dinosaur cousins into lunch and leather goods.

In Australia our **Saltwater Crocodiles** are colloquially known as **'Salties'**. They are a large opportunistic apex predator inhabiting river banks, coastal mudflats, brackish wetlands and even venturing out into coastal waters. They prefer to stick to the mainland, but have been known to swim great distances at sea and can turn up in the most unexpected places. A **Saltie** will ambush most of its prey before either drowning it or swallowing it whole. It's capable of dispatching or disabling almost any animal that enters its territory: stock animals, feral pigs and water buffalo, other apex predators such as **Bull Sharks** and **River Sharks**, as well as large **Sawfish** and **Shovel-nose Rays**, which can inhabit both freshwater and saltwater and grow to around six metres. While this may illustrate the awesome power of these creatures, their staple diet is one of fish and mud crabs. Other prey can include various reptiles, birds, small mammals and, while unprovoked attacks on humans aren't that commonplace, we are definitely on the menu.

That **Crocodile** attacks aren't as common as you might think is thanks to several factors. Firstly, there is a plentiful natural food supply; secondly, culling and management of larger males or potential rogue **Crocodiles** keeps them in check; thirdly, awareness is widespread and danger sites are well signposted. Culling used to be unnecessary as, for many decades, Europeans hunted them for their skins - nearly to extinction. Australians in the far north have now learnt to live far more in harmony with these 'prehistoric monsters'.

Worldwide, data on attacks are limited outside Australia, with only two fatal attacks reported on average per year. In Australia, from 1971 to 2013, the total number of

OPPOSITE PAGE
Lurking in the shadowy canopy of the riverbank a fearsome monster - The Australian Estuarine Saltwater Crocodile.

THIS PAGE
Saltwater Crocodile resting.

fatalities reported due to **Saltwater Crocodile** attack was 106, highlighting the potential these fearsome creatures have. Living alongside these dangerous creatures today is important for conservation, tourism and the health our northern ecosystems, but we can never be too careful.

Appearance *The Australian Saltwater Crocodile is strong and solid, seemingly untouched by evolutionary time. It has a wide snout compared to most crocodiles, though it is longer too. A pair of ridges runs from the eyes along the centre of the snout and it has fewer armour plates on its neck than other crocodiles. The scales are oval in shape and the scutes are either small compared to other species or, not uncommonly, entirely absent. In most other respects - eyes, body, build, tail and limbs - they are similar to many other species.* **Size** *Male Salties average in the wild around 5-6m long, but they can grow even larger and have been recorded over 7m (in Australia, the largest measured was 8.4m), weighing in at a around 1,000kg. Females are much smaller, growing to be about 3m long and weighing 150kg.*

Crocodiles in Australia prefer the far north. A couple of years back I visited the crocodile infested Proserpine River, an hour or so drive south from Airlie Beach and the Whitsunday Islands. There I took a safari tour and explored the southern-most reach of these crocodiles' territory. It was fascinating to see these creatures in the wild. I'd been swimming off a yacht out of Bowen, Shute Harbour and Airlie Beach only weeks before my visit. Twice I had to remain out of the water after crocodile sightings and warnings were issued. Living with these creatures is very much a real scenario up north.

The southernmost Saltwater Crocodile caught in Australia hangs over the bar in a Wonthaggi pub in Gippsland, Victoria. Discovered and shot in the 1900s at Powlett River after livestock was taken, this croc has become a great hit with tourists.

The timid Freshwater Crocodile

While Salties dominate the estuaries and coastlines of the far north of Australia, the more timid and relatively harmless Freshwater species is found more up river and inland, thriving in billabongs, streams, rivers and even lakes like man-made Lake Argle. The smaller ecosystems probably ensure these crocodiles stay smaller over time. A little like 'insular dwarfism' they have had no need to grow bigger to fill this natural niche.

The shy and secretive **Freshwater Crocodile** is found from the Kimberley to far north Queensland. Unlike the larger and more dangerous **Saltwater Crocodile**, this species is pretty timid and quick to hide when confronted by any loud human disturbances. However, swimmers cooling off can be potentially at risk from a defensive bite if they accidentally corner a submerged crocodile.

When threatened, the **Freshwater Crocodile** has some pretty unique defensive measures. Firstly, they inflate, shake and shudder their body, causing the surrounding water to ripple violently. Secondly, **Freshwater Crocodiles** also use a low-pitched warning growl. If approached too closely, these crocodiles will bite or strike out: most likely with a quick defensive bite, like a snapping dog. But these can cause serious cuts and puncture wounds.

A bite from a really large **Freshwater Crocodile** could be far more dangerous, causing serious damage or even a secondary infection from deeper puncture wounds and cuts. So while relatively harmless compared to their bigger cousins, **Freshwater Crocodiles** can defend themselves and can be dangerous.

Appearance *This crocodile's colour ranges from grey to tan-brown, with dark patches along the sides and top of the body. The nostrils and eyes sit at the top of the head and the fine sharp teeth are clearly visible even when the mouth is closed. The powerful tail features large triangular scales along its length, which is almost half the total length.* **Size** *Males can grow to 3m long, while females grow to 2m.*

- Saltwater Crocodile
- Freshwater Crocodile

THIS SPREAD
Two Egrets stalk the shallows, while two Freshwater Crocodiles beat the northern heat submerged in the cool billabong.

Great White Shark

Throughout my life I've always snorkeled, spearfished and scuba dived. One evening, close to dusk, I went abalone diving off the coast of Gippsland. As I swam out, I was joined by a couple of sea lions on the reef. It was a bucket list moment, collecting my last abalone, 4 metres down, 10 or 20 seconds into my breath and tangled in kelp. I suddenly realised the seals had disappeared. I looked around and out of the deep a Great White Shark silently cruised towards me. Honestly, for a split second I was startled, scared out of my mind, but I had few options and the shark was nearly on top of me as I clung to the kelp stalks - so I froze! It just glided on by. An overwhelming feeling of awe came over me as this majestic creature cruised on past. It was magic! It was so beautiful and the feeling of respect and calm still fills me to this day.

The **Great White Shark**, historically known in Australia as the **White Pointer**, has to be one of the world's most feared sharks. Often depicted as a ferocious man-eater, the **Great White** is without a doubt considered a highly dangerous creature and, with no known natural predators other than the killer whale, it hunts down its prey with unflinching speed and power.

Humans aren't the shark's preferred meal. The **Great White Shark** mainly eats sea lions, seals and their kin, sea birds, and large fish like Kingfish, Giant Travalley or Tuna. They will often also scavenge off whale carcasses floating in the remote oceans.

Nevertheless the **Great White Shark** is responsible for the largest number of reported unprovoked shark attacks on humans worldwide. Although these attacks are still very rare, our fear fuels shark tourism, including cage diving encounters, and our fascination has created well-known book and movie franchises such as *Jaws*.

Not a lot was known about these creatures for many years, but in recent decades research has begun to unravel the mysteries of their migratory and breeding patterns and behaviours, population size, and the hunting and feeding practices of these huge predators of our southern waters.

Australia is surrounded by 25,760 kilometres of coastline and 8,222 islands, coral reefs and shoals making up the oceanic territory. Within these coastal waters is the highest diversity of shark species on the planet. From over 400 species of shark worldwide, 170 are found in Australian waters. However, only 12 of these have been known to make unprovoked attacks on humans - the Great White Shark tops the list in terms of number of incidents.

All sharks in general get a pretty bad wrap, but when you look at the science and statistics, this is really because of only four species. Of the twelve species that have shown unprovoked aggression towards humans, the **Great White**, **Tiger Shark**, **Oceanic Whitetip Shark** and the **Bull Shark** are the only four that have been responsible for fatal attacks over the past 20 years. A handful of other species are always on our watch lists like the **Great Hammerhead** and sharks from the **Whaler** family (related to **Bull Sharks**), but the rest are either too small or too meek to warrant any real fear at all.

Australia loves its beach culture. Our coastline has roughly 12,000 pristine beaches and it's estimated as a population collectively, we take 500 million dips in the ocean a year. That's alot of activity to attract the **Great White sharks**, and plenty of activity to scare away the smaller sharks or the more timid species.

Even with a 71% decline in shark populations worldwide since the 1970s, there has been a rise in shark attacks and fatalities. In the 230 years of keeping detailed shark attack records, there have been (as of 2021) 1,068 shark attacks in Australia, with 237 of them fatal. Increasing conservation efforts and research into shark behaviour is helping us get a better understanding of these creatures. Some species are bouncing back, but statistically shark encounters are on the increase too.

There is no rhyme or reason to nature. Shark attacks occur all year round, although seasonal peak periods occur mostly between November to April. Worldwide, most shark attacks happen in the daytime, with (unsurprisingly) more on weekends during the warmer seasons. As we are all land-lubbers, even for those of us with a passion for fishing, surfing, diving and snorkeling, the ocean isn't our home.

It's great for a dip or recreation, but if you get in the ocean, you always run the slight chance of a rare encounter with a shark. You are diving into their domain. You are swimming in their fish pond, dipping into their lounge room and jumping into their favourite restaurant.

The statistics indicate three things: that Australia has a lot of coastline and loads of sharks; we have a small population that frequents the beach more than in most other countries; and we have around 4-6 shark attacks per year from which 1-2 people die. We may never solve the riddle of a **Great White's** nature and people will never stop looking for adventures in the deep ocean wilds. You will almost certainly never even see a **Great White**, but there are very good reasons why they are considered so dangerous.

Some will always find the courage and seek the excitement in facing their inner fears. Others will enjoy watching *Shark Week* on the *Discovery Channel*. Whatever way you see these ancient creatures, you can't help but admire and respect how potent and well-evolved they are. Carelessly swimming where you shouldn't might just give you that unwanted rare encounter with a **Great White Shark**, and maybe this encounter will teach you a lesson in your own foolhardiness or courage!

Appearance *The Australian Great White Shark is a large stocky fish with a body shaped like a blunt submarine. They have a sharp snout, large fins, and a strong crescent-shaped tail. Only the belly of these sharks is whitish. Distinctive menacing eyes complement the many rows of large replaceable teeth which fill their large, hugely powerful jaws.* **Size** *The largest fully grown White Sharks get to about 6.4m in length.*

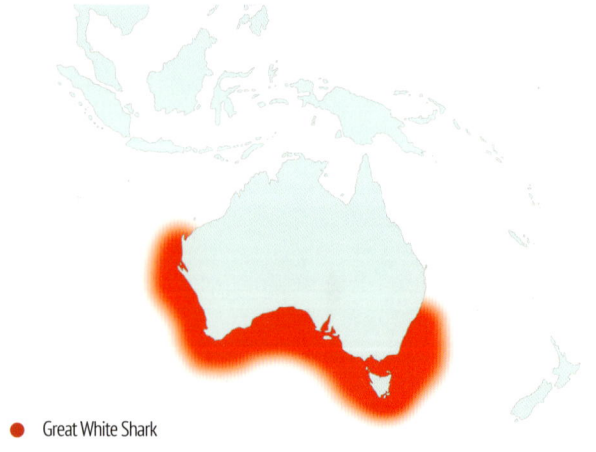

● Great White Shark

PREVIOUS SPREAD
A Great White Shark hunts five young Sea Lions in the open sea off the Tasmanian coast. Quick manoeuvrability and sufficient numbers to distract this apex predator may the help the Sea Lions escape but the shark is an expert hunter.

OPPOSITE PAGE
Majestic and calm this king of ocean, The Great White Shark, travels alone and covers great distances over the course of its lifetime.

THIS PAGE
Two Bull Sharks chase mullet in silty river water in an estuary on the New South Wales coast.

The Aggressive Bull Shark

Bull Sharks are highly unpredictable, often aggressive, and can make sudden surprise attacks on spear-fishermen. These lethal opportunists like brackish estuarine river systems, canals, marinas and man-made harbours. Adapted to traveling upstream into freshwater, and comfortable in saltwater, Bull Sharks primarily hunt bream and barramundi and would have to be one of the most dangerous sharks in this book.

Which shark is the most dangerous? The answer will always depends on who you're talking to. The 'big three', as they've come to be known, are the **Great White Shark**, the **Tiger Shark** and the **Bull Shark**, and each carry their fair share of dangerous attributes and adaptions to different ocean ecosystems. That being said, most shark experts will agree that the Bull Shark, despite not having notched up as many kills as the **Great White**, is the most unpredictable and dangerous. Its preferred hunting grounds include our marinas and areas where we swim and fish. The threat from these sharks is notable, although attacks are extremely rare. If they are known to be around, you definitely enter the water at your own risk.

With a characteristically blunt head and strong streamlined body, this shark is built for business. Known for its highly aggressive behaviour, the **Bull Shark** moves without hindrance between freshwater and saltwater environments.

These sharks are often seen cruising man-made canals or inhabiting the artificial waterways that allow the passage of boats from multi-million dollar coastal developments, with private boat sheds, to their offshore playgrounds, harbours and bays. The sharks' presence in these often murky waterways prevents any humans enjoying a dip from their private jetties.

Shute Harbour, near Airlie Beach in Queensland, is the gateway to the Whitsunday Islands. Local tourist boats come and go daily, ferrying people to the islands. Yacht and dive charter boats use the harbour and the surrounding bay, and small islands shelter many private moorings. There are many shallow reefs with deep channels and coral alleyways.

In 2017, Cyclone Debbie ripped through the Whitsundays, destroying much of the coral ecosystems and holiday resorts on the outlying Islands. The cyclone left a lot of the coral damaged, and the silt and sediment and lack of clarity in the water has taken years to clear up between the coastline and outer Great Barrier Reef.

Without coral to clean up the water and heavy rains pushing dirt and land born sediment into the reef system, the coastal area has seen a rise in crocodile and shark related reports.

With murky water comes the element of surprise and **Bull Sharks** love to hunt in these conditions. Sailing around the Whitsundays myself, I swam and spearfished these waters, which were still eerie and uneasy even a few years after the cyclone. Living up there on a yacht, and talking with the dive charter crews, you get a real sense of the reality of the danger from these creatures.

One dive charter was out by one of the islands; it had been specifically chartered to put dozens of of snorkelers in close proximity to (non-dangerous) **Black Tip Reef Sharks**. Halfway through that adventure, and with tourists in the water, the crew and holiday makers were met with an unprovoked attacked by two **Bull Sharks**. While most escaped with no injuries, it shows how dangerous and unpredictable these sharks can be.

Reports of another attack a few weeks later had operators changing locations. In Shute Harbour a spear fisherman disappeared without a trace from a reef locals call Shark Alley. Then a third attack in under a month was attributed to **Bull Sharks**, although with no eyewitness accounts the supposition was simply based on the shark's nasty reputation.

Appearance *This shark is recognisable by a combination of characteristics: a stout body, blunt snout, serrated, razor-sharp teeth and very little markings once a grown adult. It has a second dorsal, no skin ridge between the dorsal fins and small eyes.*
Size *The species grows to around 3.4m.*

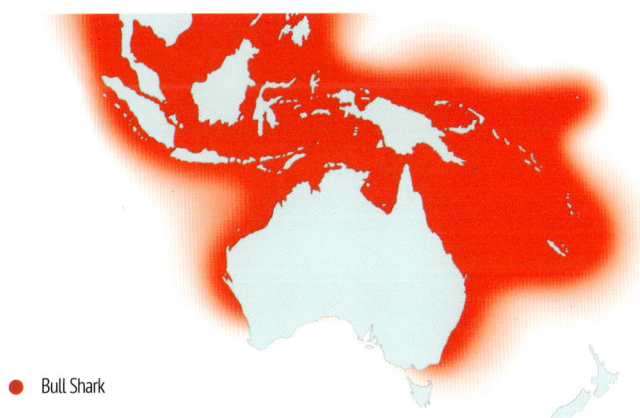

● Bull Shark

Our Fearsome Tiger Shark

Tiger Beach in the Bahamas is the place to go if you want to swim with Tiger Sharks. Following strict rules while diving here, you can safely swim without a cage. The Tiger Sharks in the YouTube films of these encounters all look placid, calm and friendly on the clear sandy shoals. Yet when you read the science and statistics on the wild Tiger Sharks in our Australian waters, there is clearly another side to them that is far more dangerous.

In my personal adventures, I have had two close encounters with these sharks. Both times I was spearfishing and both times there were nearby boats, fishing and chumming for sharks. On one occasion, after the boats left, I returned to shore across the reef. My catch and activity attracted a rather large female Tiger Shark. I calmly and promptly nudged the shark with my Hawaiian sling and she returned to the deep. In hindsight, if it hadn't been a clear day with good visibility, the story could have played out very differently.

The **Tiger Shark** gets its name from the distinctive vertical bars covering its body. These bars fade slightly as they age, while they are boldly visible on **Tiger Shark** juveniles. The females can reach five meters in length making the **Tiger Shark** the fourth largest shark in the ocean and second largest predatory shark, behind the infamous **Great White Shark.**

Shark attack statistics indicate that Australia is the deadliest location in the world when it comes to deaths directly associated with shark attacks. **Tiger sharks** are responsible for more recorded attacks on humans than any shark except the **Great White**. But, while some people regard the **Tiger Shark** as a large voracious and formidable ocean predator, others believe these sharks to be calm, friendly and curious. Surfers, snorkelers, divers and swimmers need to know about this unpredictable 'Jekyll and Hyde' behaviour: the sharks can be both calm and curious or aggressive when they approach humans in the water.

Appearance *The skin on these sharks can range from blue to light green with a white or light-yellow underbelly. This countershading camouflages the shark from prey. Dark spots and stripes are most visible in young sharks and fade as it matures.* **Size** *Adults grow up to 5m long, with females larger than the males. Mature females often go over 4m while mature males very rarely get that large. Even larger unconfirmed catches have been boasted about over the years.*

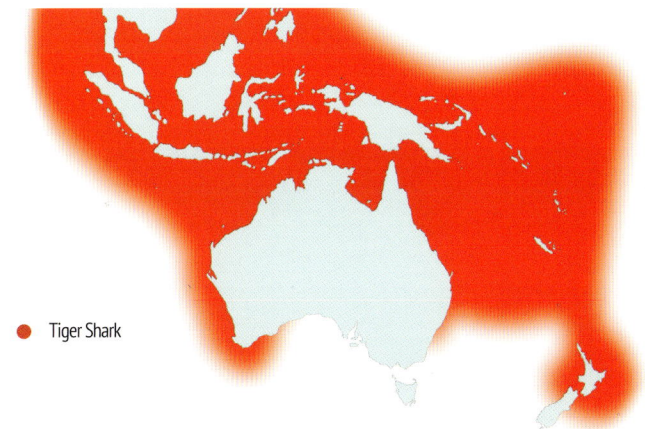

● Tiger Shark

OPPOSITE PAGE
Four Tiger Sharks patrol the outer reef system, part of the World Heritage Great Barrier Reef.

Hammerhead Sharks

My only in encounter with a Hammerhead Shark was while getting ready to go spearfish in the Whitsundays. We anchored the yacht, prepared the launch and geared up to head out across the reef. A curious Great Hammerhead came to the surface to investigate. Startling at first, the experience was heart-stopping before fascination and admiration kicked in. As quickly as the shark made its appearance, it dived away and disappeared into the depths. Undeterred, we set out to grab dinner.

Most known **Hammerhead Sharks** are small and considered harmless. However, the unpredictable nature and size of the **Great Hammerhead** makes this shark a potentially dangerous species. By all accounts, there have only been just over a dozen unprovoked attacks by **Hammerhead Sharks** on humans documented, although there have never been any known fatalities.

Hammerheads inhabit warmer waters along our coastlines and continental shelves worldwide. Different to most sharks, some **Hammerheads** often swim in schools during the day, becoming solo hunters at night. Their cephalofoil or unique Hammer-like head, gives these sharks superior binocular vision and depth perception to hunt prey.

There are nine known species of **Hammerheads** worldwide and three recorded species in Australia including the **Great Hammerhead Shark**. As they are often seen foraging in the shoals close to beaches, there is enough reason to remain cautious around these sharks.

Appearance *They have usually light gray bodies with a greenish tint, and white bellies to blend in when viewed from below. Their heads have a hammer-like shape.* **Size** *The known species range from 1m to as long as 6m.*

THIS PAGE
Two Hammerheads patrol the beach off Queensland's Gold Coast.

Pack Hunting Bronze Whaler Sharks

I have a healthy respect for Bronze Whalers. When I first encountered these sharks I was abalone diving off Gippsland, Victoria. I was returning from my dive with my catch, just on dusk in murky water filled with schools of mullet. A couple of these Whalers were hunting the mullet. They made themselves known by swimming circles around me as I crossed the reef. A few times they came in quite close, but after a slow tense swim, I returned to the shore and they returned to hunting the fish.

Whaler Sharks are known throughout the world to inhabit shallow shorelines and often utilise estuaries or travel up rivers. Included in the family are **Bull Sharks**, **Spinner Sharks**, **Dusky Sharks**, **Black Tip Reef Sharks**, **Copper Sharks** and **Bronze Whalers**.

Bronze Whaler Sharks are found in many Australian coastal habitats including bays, shoals, harbours, surf zones, shallow reefs and close to other shorelines such as beaches. They may be aggressive predators, but they are not known for their aggression towards humans and are rarely implicated in attacks.

The only known fatal attack by a **Bronze Whaler Shark** happened in September of 2011 in a shark attack hotspot: Bunker Bay, Western Australia. Records list about thirty unprovoked attacks on humans and boats, none other of which have been fatal. Attacks on humans by **Bronze Whaler Sharks** have been mostly on spear-fishers and it is thought the sharks were trying to steal their catch.

Interestingly, the **Bronze Whaler Shark** doesn't normally hunt individually, preferring to hunt prey in small packs or large groups. The numbers are used to their advantage when taking down larger prey or bait balling mullet, sardines and other schooling fish species.

Appearance *They have a slender streamlined body with a slightly arched profile, a rather long and pointed snout with the nostrils preceded by low flaps of skin.* **Size** *They reach a maximum size of 4m in length.*

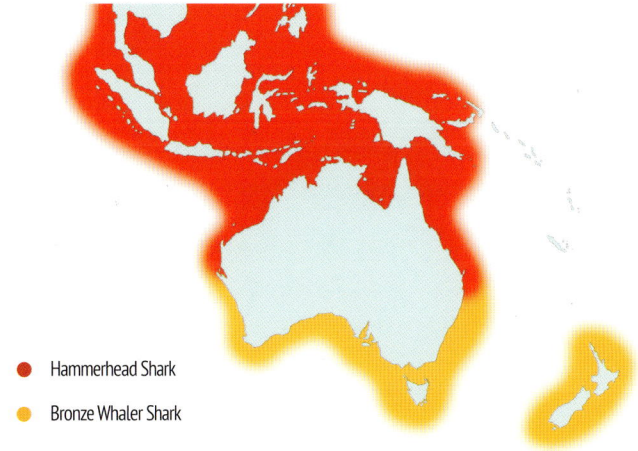

- Hammerhead Shark
- Bronze Whaler Shark

THIS PAGE
Two Bronze Whalers cruise the surface.

Smooth Stingray

Big, gentle and playful, these stingrays delight snorkelers and scuba divers (myself included) the world over. Stingray attacks are exceedingly rare, fatalities even rarer. Since 1945, in Australia, there have only been two deaths. One stingray tragically claimed the life of an Aussie legend: Steve Irwin, 'The Crocodile Hunter' and 'Wildlife Warrior'.

Globally there are over 200 stingray species inhabiting our oceans. In Australian waters there are close to 50 known species. The most common and popular with divers are the **Smooth Stingray**, **Southern Eagle Ray**, **Southern Fiddler Ray** (locally known as the **'Banjo Shark'**), and the **Coastal Stingaree**. The **Smooth Stingray** is the largest Australian stingray, growing to 4.3m in length and a weight of 350kg.

Stingrays harbour their weapons for one purpose: protection. Tail spines are an effective deterrent to predators, such as sharks who commonly eat stingrays.

They are usually non-aggressive, curious and playful creatures in the presence of divers and snorkelers. Their first instinct is to swim away if they feel threatened, but - as with all marine life - people must respect a stingray's personal space. Like all true stingrays, **Smooth Stingrays** have one venomous spine (the sting), halfway along the tail, which is capable of inflicting severe or potentially fatal wounds. It can raise its tail above its back like a scorpion and, with lightning speed, turn and flick the barb into any exposed part of your body.

Appearance The Smooth Stingray usually has irregular rows of small white spots on the upper surface beside the head. There are no thorn-like denticles along the dorsal and mid-line of the disc, as in many other stingrays. **Size** These stingrays grow to 4.3m in length (with a 2m-wide disc) and a weight of 350kg, although there are always stories of bigger Smooth Stingrays.

Personally, and this might sound like a 'tall tale' or 'fishing story', in all my snorkeling and spearfishing, the largest I've ever seen had what seemed to me a nearly 5m disc, maybe more. But that's a story for another book!

• Smooth Stingray

THIS SPREAD
On a sandy shoal two Smooth Stingrays feed.

Australia's Dangerous Cassowary

Cassowaries are potentially dangerous and can be particularly aggressive when threatened or protecting their young. Out of around 150 known attacks, the only documented human death caused by a Cassowary was on April 6, 1926. A couple of young teenage boys came across a Cassowary on their property and attempted to kill it, one of the boys coming off second best!

The **Cassowary** is a flightless bird, closely related to the emu. Although the emu is taller, the **Cassowary** is heavier, the heaviest bird in Australia and the second heaviest in the world after the ostrich. It is covered in dense, two-quilled black feathers (similar to the emu) that look more like hair. These feathers are not for flight, but for regulating body heat and temperature, as well as protection from its rainforest habitat and environment. The feathers keep the bird dry and safe from the sharp thorns found on many rainforest plants.

The **Cassowary** is Australia's most potentially deadly bird, impressive in both size and strength. They have killed before with powerful kicks and dagger-like talons which can easily disembowel a person in one swift slash. They will defend their young if needed and can become aggressive if they feel threatened. The **Cassowary** is rightfully considered the most dangerous bird in the world!

Let's put this in some detail. Those three dagger-like claws are big and sharp. Used for gathering food, scratching and digging, the inner toe has a huge 10 cm long claw. Put some force behind that and a **Cassowary** could slice you open faster than an electric can opener goes through a tin can. Powerful legs help the **Cassowary** run fast through the dense rainforest underbrush, reaching speeds of 50km per hour.

Forget superman! A **Cassowary** can jump nearly 2 metres straight up into the air and swim like *Aquaman* (well maybe not that good, but you get the idea...). This bird is very capable of fending off threats and escaping danger. On a quieter note, that long claw we mentioned is particularly handy when digging for fallen fruit in the leaf litter.

Then there is the wild head-gear. All **Cassowary** species (by the way there are three) have a 'casque', also called a helmet, that starts developing at one to two years of age. The casque is made of a sponge-like material and covered with a thick layer of keratin (the same thing our fingernails are made of). No one knows for certain why **Cassowaries** have a casque. It could be for courtship, male dominance, attacks or just protecting the bird's head as it pushes through the heavy rainforest.

While the casque mystery remains, we can all agree it's an iconic warrior-style headdress that gives these dangerous bush creatures a unique mystic as they forage through the deep rainforests in search for fruit.

Appearance *Adults have generally jet black feathers, their face and neck varying according to species and location. Vibrant blue/pink/red skin distinctively stands out against their robust bodies. Female Cassowaries are larger than the males and are even more brightly colored. All three species of cassowary have casques and they are a major feature of these iconic flightless birds.* **Size** *Among the largest birds on the planet, the Southern Cassowary is the largest, females pushing 2m in height. Males weigh up to 55kg and females are slightly heavier.*

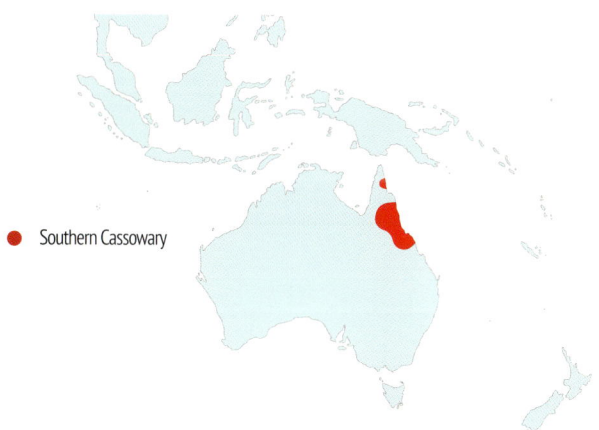

• Southern Cassowary

OPPOSITE PAGE
Up-close and personal with Australia's largest flightless bird.

Swooping Magpies

In my garden at home I often have Magpies nesting. My resident Magpies have never swooped me and they seem quite content. In 2021 a large storm hit Melbourne and blew their nest out of the tree and my Magpies let me re-establish a new nest to keep cats and foxes from preying on their youngsters. In my experiences with birds, I have rarely been attacked or swooped. But in Australia swooping native birds that nest in urban areas can be a real problem and have caused panic, fear and some nasty injuries.

There are few birds that are as familiar to Australians as the **Australian Magpie**. They are pretty common in urban areas, pied (striking black-and-white) medium-sized ground-feeding birds. They hunt manicured backyard lawns for worms and insects, stalk open grassland and council parks and nest in gardens and tree-lined streets. This close proximity our urban life means **Magpies** live within the orbit of our everyday activities and put up with our hustle and bustle.

Most **Magpies** don't actually see humans as a threat and are really unlikely to swoop at all. But in recent years some **Magpies** have turned this protective swooping behaviour onto human 'threats' and swooping **Magpies**, with their sharp beaks, strong robust bodies and lack of fear of humans, are now a risk to public health. A perhaps over-protective culture has seen local councils putting up cautionary signs and warning the public across Australia.

Some people have a genuine fear of swooping birds. They can definitely startle you! And, if a beak does connect with your skull with any force, it can be very painful, even potentially cause some damage and an immediate trip to the hospital.

Magpies are a widespread and quintessential Aussie icon. So their dive-bombing has many people concerned, even a little panicked. Kids walking to school, oblivious joggers with ear buds in, bicycle riders and pedestrians across Australia are growing more and more fearful of these black-and-white aerial attackers.

There isn't much we can do about these birds: it's often a temporary behaviour and only seasonal. Many people in urban areas are modifying their behaviour to accommodate these much loved bush creatures. Dangerous or not, as the sun comes up Aussies still love the morning chorus from our beloved **Magpies**.

● Australian Magpie

OPPOSITE PAGE
Look Up! Look out! Here come the swooping Magpies, aerial dive-bombers with strong sharp beaks that are potentially dangerous.

THIS PAGE
A large species of Butcherbird, these two Magpies spend most of their days stalking lawns and gardens for worms and insects.

Our Iconic Big Red Kangaroo

Traveling at night on a dark desert highway in Australia is unwise and dangerous. All manner of animals - stock such as cattle and sheep, feral animals like foxes and goats, and of course native animals - forage by the roadside and wander onto the road. Grazing kangaroos are no exception. Dazed by headlights, they flirt with death by bounding out unexpectedly and often collide with vehicles, causing damage and injuries. Apart from collisions, male 'Big Reds' are very intimidating and they could easily tear you apart if provoked. Once while driving to Broken Hill, at 2 am, I saw a monster Big Red, possibly 2.5 metres tall - humans are no match for that size kangaroo.

The **Red Kangaroos**, affectionately known in Australia as **Big Reds**, are the largest of all kangaroos and are found in abundant numbers across our arid interior. They have only benefited from the spread of agriculture, grassland for cattle, irrigation channels and man-made waterholes.

Because kangaroos are usually docile, attacks or violent encounters with people are rare. However, they can attack when provoked. They especially target dogs and have been known to kill domestic pets. It's something to keep in mind, especially with the huge size and bulk of these **Big Reds**.

Their antipathy to dogs has probably have arisen over the last few thousands of years since Dingoes were introduced to Australia, the defense mechanism perhaps triggered by a faint genetic memory of being hunted by *Thylacines* and *Thylacoleos*.

Kangaroo attacks have been on the increase in recent years, as they have moved into parklands and become more commonplace in urban areas. In the 2000s attacks have increased on average from 5 a year to reportedly a dozen or more. In saying this, there is only one recorded human death attributed to a kangaroo in our history. In the 1930s a hunter attempted to save his dogs and was killed.

Collisions remain the big danger from a **Big Red**. Their bulky size severely damages vehicles and can completely write-off smaller cars. The risk of injuries to occupants is greatly increased if the windscreen is the point of impact. Hence why kangaroo warning signs are commonplace throughout Australia. Smaller kangaroos, like **Eastern Greys**, cause less damage, but a **Big Red** impact at speed can be fatal. On average, 18 Australians die in car accidents related to kangaroo impacts every year. The most dangerous time is when they are roadside feeding, between 5 and 10 pm particularly after extended dry-spells.

Appearance *This, the largest kangaroo, has long, pointed ears and a square shaped muzzle, with short, red-brown fur, fading to pale buff below and on the limbs. Females are blue-grey with a brown tinge, pale grey below, although arid zone females are coloured more like males. It has two forelimbs, with small claws, and two muscular hind-limbs, which are used for jumping and fighting. It has a strong tail which is often used to balance when standing upright.* **Size** *Males grow to a head-and-body length of 1.3–1.6m with the tail adding 1.2m. Long legs can add nearly another metre to the standing height. The largest officially recorded Big Red stood a bit over 2m. However, in the deep remote deserts, sightings of much larger Big Reds are rumoured. Females are considerably smaller.*

● Red Kangaroo

THIS SPREAD
Big Red Kangaroo bounds across the desert.

Australia's Wild Dingoes

I'll never forget meeting my first wild Dingo. What an experience!
I was traveling through the MacDonnell Ranges, Tjoritja, in Arrernte country, returning to Alice Springs, when I saw this large russet-furred Dingo, trekking along the roadside. Pulling over, to my surprise it curiously stopped and watched me. Cautiously, I stepped out of the vehicle and took a few pictures. The Dingo kept its distance and so did I. But there was this silent exchange going on between us while I admired this beautiful bush creature. Suddenly, in the blink of an eye, it disappeared into the scrub like a ghost.

Dingoes hold a significant place in the spiritual and cultural practices of Indigenous groups and Torres Strait Islanders. They feature in dreamtime stories, ceremonies, on cave paintings and rock carvings. They are found across most of the mainland, from deserts to snow-covered alpine areas, from grasslands to rainforests, where they often dominate as a keystone species. However they seem to favour edges of forests skirting open grasslands. They usually remain in one area, around a central den, but can cover large distances beyond their territory if needed.

They mainly hunt alone at dawn, dusk and during the night, communicating with wolf-like howls, instead of barking. The reason why we should consider these animals as dangerous is displayed in their hunting skills. Depending on the size of their prey, they often work together in a cooperative pack. Taking down large kangaroos and even fearlessly tackling feral water buffalo.

Wild **Dingoes** generally avoid humans, having a natural wariness of us. **Dingo** attacks have mostly involved people feeding them, particularly on K'gari (Fraser Island) where 4 wheel driving, ecotourism and dingo-related tourism bring Dingoes and humans together. Dingoes have been interacting with campers for decades.

Due to increasing aggression towards humans, National Parks selectively culled the Fraser Island **Dingo** population curbing the problem.

On the Mainland the vast majority of attacks are minor, few have been fatal. The most famous **Dingo** attack being the death of Azaria Chamberlain in the Northern Territory during the 1980s.

While many National Parks in Australia have good signage advising visitors not to feed **Dingoes**, people still continue to feed them or unwittingly encourage **Dingo** interactions.

• Dingo

OPPOSITE PAGE
Yawning Dingo resting in open grassland.

THIS PAGE
K'gari (Fraser Island), Dingoes beach-combing for washed up remains are unphased by the 4WD tracks.

Our Beloved Platypus

In all my years I have only sighted a Platypus twice in the wild. They are quite elusive creatures and spend a lot of their time in burrows or under the water diving for yabbies. I spotted my first one, just briefly, on the surface of a dam at dusk, and saw another in a small creek while hiking in Victoria. So why would such a cute and elusive bush creature be considered dangerous?

The **Platypus**, or **"Duck-billed" Platypus**, has to be one of Australia's most adored and iconic bush creatures. They truly look like something out of the dreamtime. Shy and elusive, they inhabit healthy freshwater streams and waterways of eastern Australia from Queensland to Tasmania. Yet this monotreme, sharing ancestry with the Echidna, has one dangerous little secret up its sleeve - a venomous spur!

This defensive, venomous foot-spur isn't actually lethal to humans, and only the male **Platypus** produces the toxic venom, which is more painful than dangerous. The wound that the male **Platypus** spur inflicts can be excruciatingly painful and can create complications or infection if not treated. The spur is the perfect defense system and the venom can kill a **Dingo** or wild dog. It can be easily be employed on any humans stupid enough to catch and handle the **Platypus** without training or care.

Appearance The Platypus has a broad body, flat tail and is covered with dense, brown, biofluorescent fur that traps a layer of insulating air to keep the animal warm. The fur is waterproof and, with it's duck-like bill, webbed feet and eyes and ears located in a groove that closes when swimming underwater, the Platypus is perfectly adapted to hunting under water. *Size* Males are larger than the females, growing to around 50cm, whereas the females only reach around 40-43cm.

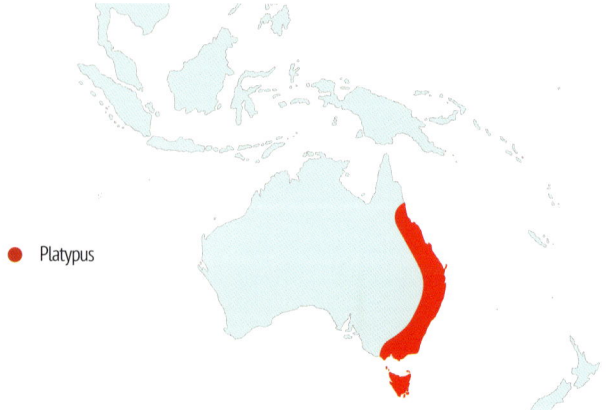

● Platypus

THIS SPREAD
Two beautiful Platypuses hunting yabbies in a clean mountain creek.

Funnel Web Spider

I was always taught to check my shoes for Funnel Web Spiders. Growing up in Melbourne I thought it was only a Sydney problem, since the full name was the Sydney Funnel Web Spider. Later I learned there are 40 species of Funnel Webs distributed from Adelaide to Sydney.

Funnel Web Spiders are medium to large spiders, their body lengths varying from 1-5cm. They typically build silk-lined burrows with open "funnel" entrances from which they get their name. The long-lived female **Funnel Web** spends most of her time in her silk-lined nest. When potential prey walks across the web, she rushes out, subduing the prey by injecting venom with her large fangs. The venom is highly toxic to humans, but does not affect the nervous system of other mammals. The onset of symptoms after a **Funnel Web** bite is less than one hour.

Around 30-40 people are bitten by **Funnel Web Spiders** each year, with 13 recorded deaths. One of these was a small child who died within 15 minutes of a bite from a **Funnel Web**.

Size *The average leg length for common Sydney Funnel Web Spider is 6-7cm.*

Redback Spider

Redbacks prey on small insects but, with extremely strong sticky webs, larger animals, such as cockroaches, spiders, small skinks and sometimes even small snakes and mice, can become entangled.

Redback Spiders, also know as **Black Widows** (because after mating the females eat the males), have a big bad reputation for such a little animal. Bites occur mostly over summer and more than 250 people receive antivenom each year across Australia. Only the female bite is dangerous, causing severe pain, sweating, muscular weakness, nausea and vomiting. In the past they have caused deaths. However, since **Redback Spiders** rarely leave their webs, humans are not likely to be bitten unless they put their hand in a web. In 2016, a bushwalker reportedly died from a **Redback Spider** bite, but this seems to have been the only death due to a **Redback** since the introduction of antivenom.

- 🔴 Redback Spider
- 🟡 Funnel Web Spider

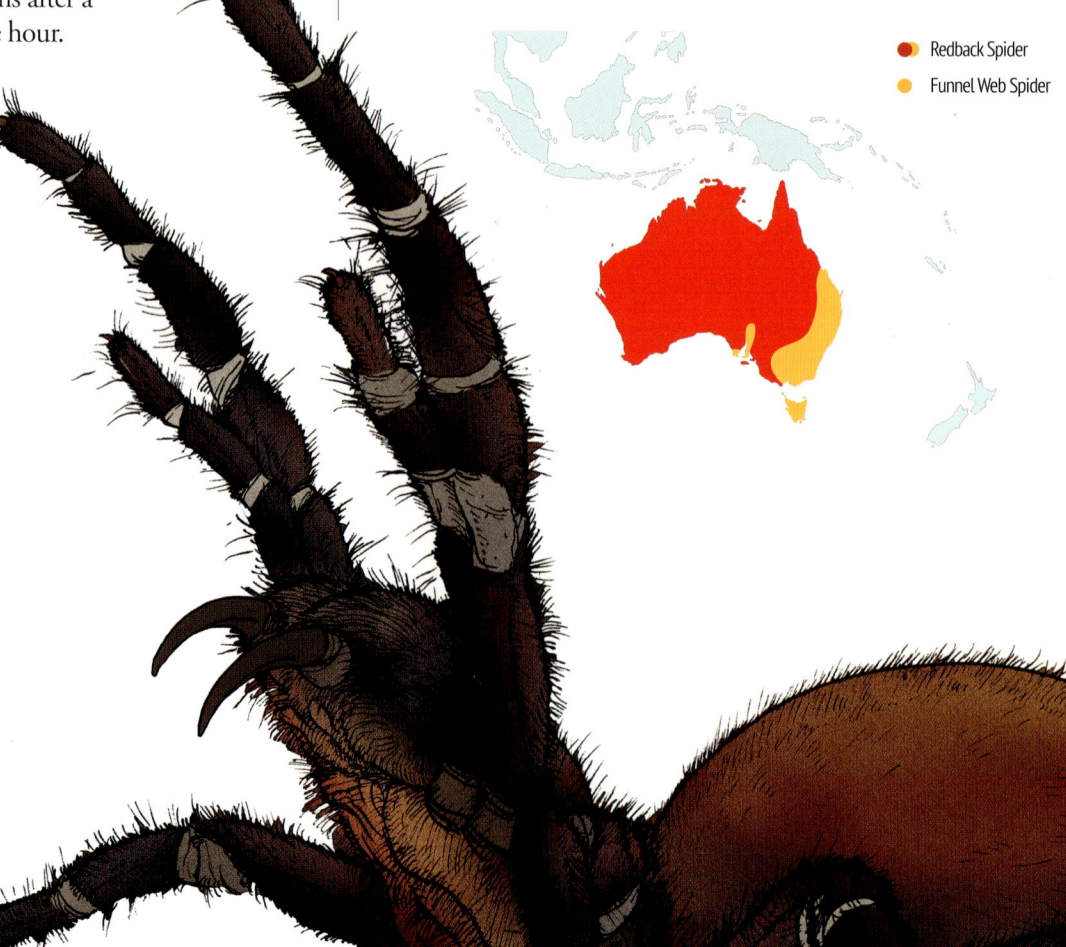

OPPOSITE PAGE
A Redback Spider secures his prey, a small skink.

THIS PAGE
The scary Funnel Web spider.

Huntsman Spider

Huntsmans aren't dangerous! So why are they in a book on Australia's Dangerous Bush Creatures? Well, while a Huntsman's bite isn't life threatening, their habit of hiding where you least expect them certainly can be! Finding their way into our motor vehicles, tucked in car doors, air conditioning vents and above us behind the sun visor can make for a pretty nasty surprise if they pop out - especially while speeding down the highway. Often growing to the size of an outstretched hand, they instill fear in most of us. Add in that element of unexpected discovery and these spiders are the stuff of nightmares!

Like most spiders, **Huntsmans** use venom to immobilize prey. They have been known to inflict serious defensive bites on humans and, when threated, rear up in an aggressive defensive behaviour. These guys are considered 'nightmare fuel' and, if social media is anything to go by, they are probably the reason many Arachnophobic travelers give Australia a miss.

Although it isn't lethal, a **Huntsman's** bite can cause localised swelling and pain, nausea, headache, vomiting, irregular pulse rate and heart palpitations, indicating some systemic neurological toxin effects, especially when the bites are severe or repeated. However, the effect of **Huntsman** bites is fraught with complications. A bite does not usually require hospitalisation, but a psychosomatic fear of them can lead lead to additional symptoms, such as hyperventilating and shock, caused simply by the high level of anxiety of being bitten.

These spiders are actually a very important predator within their habitats, with a good appetite for all those creepy crawlies you don't want in your house or home. They will happily dispatch cockroaches, beetles, other spiders, flies, centipedes, small unwanted bugs, even larger insects like grasshoppers. So swallow your fears and try to see the good in these intriguing bush creatures.

Size *Including their leg-span, a Huntsman can grow to around 15cm across.*

White-Tailed Spider

White-tailed Spiders are one of Australia's most feared spiders. Reports of bites turning into necrotic ulcers have become the stuff of legend. Like a good zombie movie, fueled by over-active imaginations, horrific ideas and myths of incurable ulcers, rotting flesh and blood infections still abound today.

White-tailed Spiders are native to southern and eastern Australia, and so named because of the white tips at the end of their abdomens. They are wandering hunters, seeking out their prey - typically other spiders - rather than spinning a web to capture them. Reportedly, bites on humans by **White-tailed Spiders** usually display the following effects: a red mark, localised itchiness, swelling and mild pain. On extremely rare occasions bites can cause nausea, vomiting, unease and headaches. Although spider bites that create ulcers and necrosis have been attributed to these **White-tailed Spiders**, a scientific study showed these effects were caused mostly by resultant infections and not the spider's venom. A further study of 130 **White-tailed Spider** bites found no evidence at all of necrotic ulcers or infections, but this doesn't dispel the circulating myths!

Paralysis Tick

The **Australian Paralysis Tick** is one of about 75 species in the **Tick** family. It is found in habitat with high rainfall, such as wet sclerophyll forest and the 20-kilometre wide rainforest which stretches from Gippsland in Victoria and follows the eastern coastline north into New South Wales.

Most tick species' bites are harmless or only demonstrate mild effects. But on rare occasions you can get bitten by the more dangerous variety. A **Paralysis Tick** bite causes an itchy hard lump. The **Tick** will have to be removed, but sometimes symptoms don't present themselves straight away whilst the tick engorges itself. These include flu like fever, rashes, unsteadiness, weak limbs and partial facial paralysis.

Another problem associated with **Paralysis Tick** bites is that of an allergic reaction. This can range from mild itching and swelling to potentially life-threatening anaphylactic shock. In extreme cases this can lead to hospitalisation and death: there were 20 reported deaths from tick bites before 1945.

OPPOSITE PAGE
The Australian Huntsman killing a grasshopper.

THIS PAGE
Above the infamous White-tailed Spider and several examples of the Paralysis Tick, the largest one an engorged, swollen female.

Ticks are a fact of life when camping or hiking in eastern Australian. Hatching from eggs laid on thick leafy shrubs like Tea Tree, the larvae latch on to passing mammals. Even people working in their gardens can have these blood-sucking ticks burrow in for a feast. On very rare occasions a Paralysis Tick bite can infect the victim with Lymes Disease, which can kill if left untreated.

- Hunstman Spider/ White-tailed Spider
- Paralysis tick

Dangerous Insects and other Creepy Crawlies

Some of our most dangerous creatures are tiny!
Creepy Crawlies - insects and other invertebrates - can make your day uncomfortable, inflect extreme pain or just ruin your best planned outing. Sometimes it's the little things in life that can have some of the most dangerous and dramatic impacts. So let's have a look at the little beasties that make our skin crawl.

Retaining their dangerous wasp-like charactistics, **Bull Ants** or **Bulldog Ants** are large ants, growing up to 40mm long, with deep red/orange colours on the head and thorax, changing to a reddish black abdomen. They have pretty nasty pincers, slender mandibles filled with a potent venom-loaded sting. Yes, those huge choppers are pretty dangerous! If you ever get bitten by an **Australian Bull Ant** you will definitely know it and will always remember the experience. That venom can leave a very nasty welt, and it can take days for the pain to go away. If you have an Anaphylaxis reaction, enough bites can be fatal, it's best practice to avoid these guys when outdoors.

Bull Ants are everywhere. Sorry to say this, but it's estimated there are around 90 species of **Bull Ant** across Australia, with diverse behaviours and life cycles. As always there may be more as yet undiscovered, so keep a lookout.

If you think **Bull Ants** are bad news, welcome to the world of pain which is **Jack Jumper Ants**! I call them **Jumping Jack Ants**, but you aren't going to quibble about names if you're stung by these miniature mercenaries. They're easy to spot as their bodies are all black with light brown, orange to yellowish mandibles out front, and they often display jerky, jumping movements as they navigate the leaf-litter.

Most Australian native stinging ants are broadly subdivided into **Jack Jumper Ants** and **Bull Dog Ants**, which together make up the *Myrmecia genus* (group). **Jack Jumper Ants** are generally smaller, 10-15mm long. Don't let size fool you, however! Their venom charged mandibles pack a punch and they sting rather than bite. In form, the *Myrmecia* are the closest ants to wasps (all ants' ancestors). Their wasp-like stings can be very painful indeed and large local swellings can also occur, which may last a few days. Allergic reactions are not uncommon, Anaphylaxis being the most severe.

Jack Jumper Ants are widespread and live in underground nests in right across Australia, excepting the Northern Territory and the naorthern parts of Queensland and Western Australia. Some **Jack Jumper** species are highly aggressive and launch themselves by actually jumping toward intruders, hence their name.

Not native, but here to stay! **Fire Ants** have been in Australia since 2001 and are considered a "super pest". These South American immigrants affect over 50 different agricultural crops in Australia and can kill humans. It's early days, but **Fire Ants** are very dangerous and have the potential to spread to large areas of Australia, severely damaging our environment, our outdoor lifestyle, agriculture and tourism. They horrifically swarm their victims, inflicting a painfully fiery sting which can, in rare cases, cause an acute allergic reaction.

Not deadly, but looking scary, the **Australian Scorpion** can be found under moist leaf-litter, rotting logs and loose rocks. I've often discovered them under the groundsheets and tents while packing up my camp after our family holiday. All **Scorpions** worldwide possess a venomous stinger and our Australian species might be small, but it can inflict a painful sting. This results in swelling, lasting for several hours, even days. There aren't any confirmed deaths from **Australian Scorpions**, but as with all these dangerous beasties, medical advice should be sought if you are stung.

The most dangerous and deadly insect on our list is the annoying **Mosquito**! It may only live for a few weeks, but in a mosquito's brief life cycle it can cause a lot of damage, inflicting painful itchy bites and potentially spreading dangerous diseases. It's all in the **Mosquito's** saliva. Instantly causing an itchy rash, the saliva can also carry and infect a victim with parasitic diseases such as malaria and filariasis, and arboviral diseases such as yellow fever, Chikungunya, West Nile (or dengue) fever and Zika fever. Worldwide research indicates that, via transmitted disease, **Mosquitoes** cause more human deaths than any other animal: Globally over 700,000 each year. I now know why I totally dislike these little suckers!

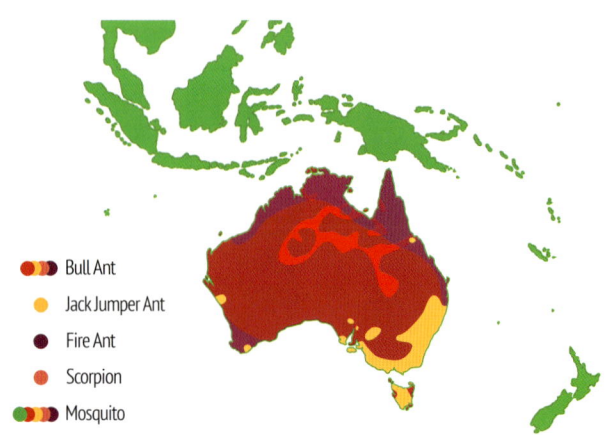

THIS PAGE
1. Bull Ants
2. Jack Jumper Ants
3. Invasive Fire Ants
4. Australian Scorpion
5. Mosquitoes

The bold Butterfly Cod or Lionfish

Apparently these fish are really nice to catch and eat. Perhaps that's because they often hunt the same fish as Red Snapper and Mangrove Jacks, which are really delicious. Butterfly Cod hunt shallow coral reefs and use their butterfly fins to startle and block smaller fish. In the brief moment that the prey need to change direction, the Lionfish snaps them up and eats them whole.

The **Butterfly Cod** or **Red Lionfish** is a venomous coral reef fish decorated with white and red/maroon/brown stripes. Adults can grow as large as 47cm in length, making it one of the largest species of **Lionfish** in the ocean. They are part of the **Scorpionfish** family and not unlike the **Stonefish**. All these fish are found in Australia and have large, venomous spines that protrude from their bodies. When swimming across the reef these long protective spines radiate out like butterfly wings, or alternatively their look can be quite similar to a lion's mane, hence the common name of **Lionfish**.

The venomous spines and dazzling display of colourful stripes make the fish seem inedible and also trick potential predators into thinking it's poisonous to eat. The venomous dorsal spines are used purely for defense. Although its sting is usually not fatal to humans, people stung by this fish will experience extreme pain, headaches, vomiting and difficulty breathing. Medical attention is strongly recommended as some people are more sensitive or allergic to the venom than others.

The **Lionfish** themselves are voracious feeders. When hunting, they corner prey using their large fins, then use their quick reflexes to swallow the prey whole. They hunt primarily from late afternoon to dawn.

Butterfly Cod are not aggressive towards people and they don't try to use their long venomous spines to impale possible threats. These are purely defensive and are mainly for show and protection. The fish are not too bothered with divers and human encounters, but avoidance is advised, so as not to put yourself at risk. Although highly venomous, they are beautiful creatures and if you leave them alone, they will leave you alone.

Appearance *Butterfly Cod are spectacular fish with brilliant colouring and long defensive spiny fins. They are part of the Scorpionfish family but, unlike other members of this family, which often rely on camouflage for protection, they swim freely around coral reefs, leaving their venomous fins to give them 360° protection.* **Size** *Adults grow as large as 47cm in length.*

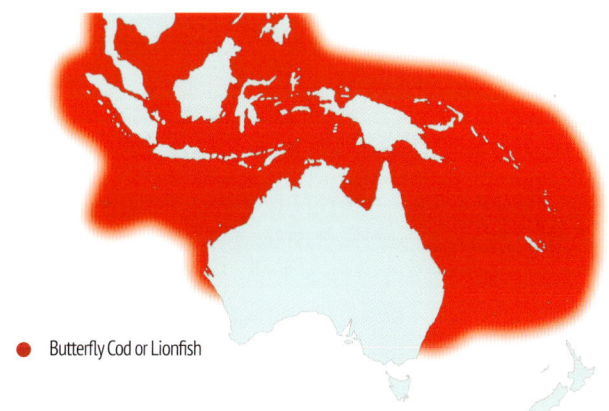

● Butterfly Cod or Lionfish

OPPOSITE PAGE
Hunting a shallow reef, the Butterfly Cod or Red Lionfish is looking for lunch.

Australia's Venomous Stonefish.

From the world's largest fringing reef, Ningaloo Reef on the west coast, across the northern waters of Australia, all the way to the Great Barrier Reef on the east coast, you may find the Stonefish. Highly camouflaged ambush predators, looking like small rock or coral boulders, they sit motionless for hours on end. They live in estuarine brackish river mouths and shallow tidal flats, half-buried in mud, silt and sand - in a rockpool or on a tidal reef-bed they blend in perfectly. Unflinching and unseen, with their strange coral-encrusted bodies, they lie on the bottom like dangerous booby traps. An unmindful beach-goer, carelessly wading through rockpools, can accidentally step on a Stonefish, encountering some of its 13 venomous spines.

The **Stonefish** is the most venomous fish in the world, but with no deaths in Australia for two centuries you could say its infamy is all just sensationalism to fill books on dangerous creatures. However, **Stonefish** haven't gone away, they haven't gone extinct: they might be elusive and hard to spot, but we still have to be cautious and heed the very real warnings. What makes the **Stonefish** so dangerous is the venom. Stored in glands sleeved around the dorsal (back) spines, this nasty toxin is delivered by when pressure-triggered or something impales on a spine. The stings induce intense pain, respiratory weakness, damage to the cardiovascular system, convulsions, paralysis and, in severe cases, death.

The local Indigenous groups of the far north have much lore and many dreamtime stories warning of the dangers of these fish. While Pacific and Indian Ocean islanders have died from stings from these creatures in recent decades, in Australia we have luckily had no recorded deaths from **Stonefish** since the first European settlements.

This might be due to the remoteness of their habitat, years of oral warnings, education, signs in populated areas or the fact that these creatures are actually quite rare and hard to find. Nonetheless they are still there, in shallow pools, estuaries and reefs, hidden and unmoving. You should still walk with caution, wear shoes and curb childish exuberance while exploring our northern beaches and rockpools.

There are two kinds of **Stonefish** in Australia, the **Reef Stonefish** and the **Estuarine Stonefish**. Both **Stonefish** are close relatives of **Scorpionfish** and **Butterfly Cod** and both are the masters of disguise! Possibly the most patient of all underwater predators, they lie in wait for passing prey. Then, with lighting speed, they suck down smaller fish, squid and crustaceans such as shrimps and crabs.

Lying perfectly still and with that incredible camouflage, you may think **Stonefish** are also very safe, but they are preyed upon too. **Sea Snakes**, sharks and stingrays all enjoy these rock-like fish. Even their co-masters of camouflage, the large **Sand Octopus** and **Reef Octopus** will seek out **Stonefish** for an easy meal, navigating the venomous spines with ease and dispatching the fish with murderous efficiency.

When you study these creatures in the wild you discover so many things you don't usually see in books. We may think dorsal spines are only good for stopping being stepped on or deterring the jaws of a reef shark or stingray. Actually its large thick muscular front pelvic fins, that fan out on the sand or reef, enable it to leap up and launch at prey or a threat within 0.015 of a second! So if you think you can handle provoking a **Stonefish**, expect a lightning fast reaction as it aggressively lurches, twists and drives home those 13 dorsal spines if you dangerously let your guard down.

Appearance *It has the uncanny ability to morph into various rock-like shapes, trying to appear like algae-encrusted rock or coral.* **Size** *Adults reach 30-40cm in length.*

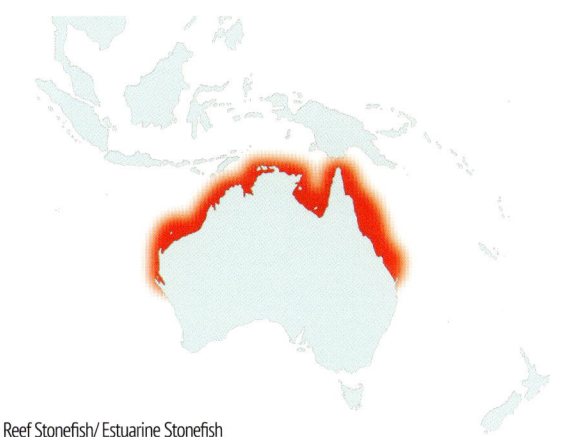

● Reef Stonefish/ Estuarine Stonefish

OPPOSITE PAGE
Seagulls squawk and squabble in the shallows, enjoying the reef at low tide. Some fish are 'crèched', protected from larger ocean predators in this sandy tidal pool. Or are they?

Two Estuarine Stonefish await their next meal, patiently waiting for the perfect moment to strike.

Cone Shells and Sea Urchins

As a kid I loved beachcombing in summer, collecting a treasure trove of washed up ocean trinkets: Cuttlefish, Paper Nautilus, Sea Urchin remains and many different shells. I was sternly warned about Cone Shells, in case an animal was still inside. Thankfully, I never come across one, but I was put in hospital by a Sea Urchin. A spine drove deep into my knuckle while snorkeling and I had to have it removed after my hand became infected and swollen. That was just a non-venomous Sea Urchin, so I can attest to the painful inconvenience of even the less dangerous of these creatures.

While **Sea Urchins** and **Cone Shells** won't attack humans they are well defended creatures. Both are venomous and there are many different varieties with various degrees of reactions to the venoms and toxins.

While **Sea Urchin** venom is mild and only produces pain, swelling and discomfort in most people, the risk of having an adverse or anaphylactic reaction from a **Sea Urchin** sting is increasing and should be something to consider and avoid.

Some **Sea Urchins** are very edible and are collected for their roe (urchin eggs). **Sea Urchin** numbers are definitely on the rise, as international ships coming and going from our shores have brought unwanted guests with them. Both circumstances have bought humans and **Sea Urchins** closer together and that puts them on our dangerous list.

Some of these these new invasive **Urchin** species have a higher toxicity and more painful sting side-effects than our native species. That **Sea Urchins** were not an issue before 1960 says a lot about our changing ocean ecosystems. **Sea Urchins** aren't dangerous to human life, but if you are stung and have difficulty breathing, lose consciousness, suffer from dizziness, mood swings or sudden changes in behavior, or get a rash or extensive swelling near the sting site, you should seek some medical attention.

It may not just be the venom that causes issues. In many cases, it can also be marine bacteria that create uncomfortable complications from encountering these spiky little balls of pain.

Cone Shells are different. They actually shoot out poison 'darts' and harpoon their victims. If you are walking barefoot through a rockpool or shallow beach surf, or sticking an unprotected hand blindly under a rock ledge or into coral, you are asking for trouble. Unfortunately our lack of awareness of our surroundings, natural curiosity and, sometimes, blind stupidity can trigger the ambush behaviour of these usually unassuming marine snails.

Although rarely seen in the wild, their decorative shells are highly prized by collectors and their venom is being collected for pharmaceutical uses. **Cone Shells** are fascinating creatures. Ambush predators, they harpoon fish, other snails, crustaceans and other small ocean creatures. Once dispatched their mouth tube stretches over the whole prey - feasting on their catch they slowly digest the remains.

Some 40 people have died from **Cone Shell** venom over the past 90 years. Even if you survive, you wouldn't want to experience the pain the venom can cause: the nasty effects include muscle paralysis, blurry or impaired vision and respiratory failure. This is why these marine snails are high on our dangerous creatures list.

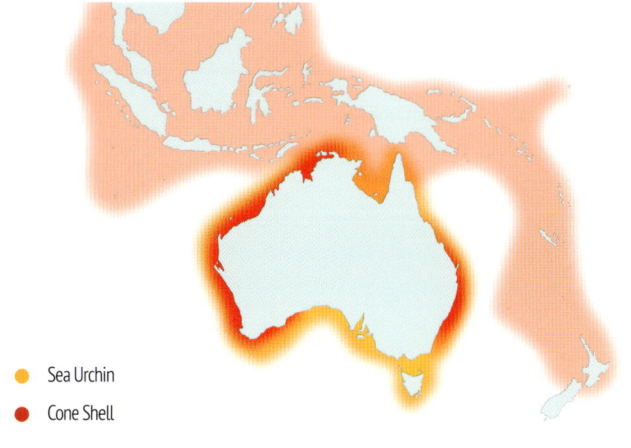

- Sea Urchin
- Cone Shell

OPPOSITE PAGE
Top; Black Sea Urchins nestle in cervices, scattered in many rockpools and reef systems across Australia. They feed at night and have conspicuous spiny defenses.

Bottom; Stalking the stands of coral, a painted warrior with blow-dart weaponry disables a small fish. Its long tubular mouth extends out to soon envelope the fish whole when digestion will begin. An infamous creature, highly venomous, the Cone Shell is highly toxic to humans and has caused deaths in the past.

Australia's Deadly Blue Bottles and Jellyfish

Camping around the coast, and a passionate snorkeler, I love to get in the water on a hot balmy day. Across south-eastern Australia, a change in the wind and weather can bring an army of little blue 'Man-o-Wars' or 'Blue Bottles'. In Australia these tiny balloon-like marauders wash up on our beaches for a few days at a time and, although they can't kill us, their their stings are extremely painful and the tentacles can leave nasty scars.

Found in the Tasman sea and washing up in summer on the New South Wales and Victorian coastlines between Sydney and Gippsland, **Blue Bottles** are the only jellyfish-like animal to watch out for in these waters. With their translucent bodies, **Blue Bottles** are difficult to spot in the water, stinging tens of thousands of swimmers in Australia every year. Though very painful, the stings aren't fatal and don't usually cause any serious complications. **Blue Bottles** are not true jellyfish and each 'animal' is actually a colony of tiny polyps, together known as a *siphonophore*.

There are plenty of dangerous true jellyfish, however! More than 2,000 different types of jellyfish have been identified around the world, but scientists believe that there could be as many as 300,000 different species yet to be discovered floating around our vast oceans. Of the 2,000 known species, 70 are identified as harmful or dangerous to humans, so who knows how many more dangerous ones there could be?

Despite their fearsome reputation, most jellyfish are completely harmless to humans. However, in the northern parts of Australia two deadly species do invade our seaside playgrounds during the summer when the waters warm up to the right temperature and the currents bring in these dangerous creatures. This is the story of Australia's notorious **Box Jellyfish**, *Chironex fleckeri*, and the **Irukandji**.

So let's get entangled in some science history. In Australia, around the 1940s, a man by the name of Hugo Flecker worked on various venomous animal species and poisonous plants. Concerned at the time with some unexplained deaths of swimmers, he identified the cause to be a seasonal visitor, the **Box Jellyfish** (*Chironex fleckeri*). In 1945 he described an 'Irukandji Syndrome' which he associated with another jellyfish. This tiny creature was later formally named as *Carukia barnesi* and is now commonly referred to as the **Irukandji Jellyfish**.

If you visit the northern coasts of Australia, you may discover that the Indigenous people knew about these dangerous creatures long before science caught up and classified them. There are many signposts, stories and warnings within their oral histories. We can never know how many fatalities came before modern records began, but fatalities are mostly caused by the larger **Box Jellyfish**, also known as the **Sea Wasp**.

One of the world's most venomous creatures, this **Box Jellyfish** entangles its victims in tentacles, causing painful welts and blistered skin. In severe cases these stingers inject a venom which causes cardiac arrest - a heart attack which can easily kill, especially when in the water.

The **Australian Box Jellyfish** caused at least 79 deaths from the first report in 1883 to when I'm writing this in 2021. While most recent deaths in Australia have been those of children, because of their smaller body mass, adults can also die. A 17-year-old boy sadly died about 10 days after being stung while swimming at a beach on Queensland's western Cape York in February 2021.

While only two deaths in Australia have been attributed to the **Irukandji Jellyfish**, people stung by these tiny thumbnail-sized creatures suffer not only intense physical pain, but also psychological symptoms, still known as the Irukandji Syndrome. So the **Irukandji** is definitely considered an extremely venomous and highly dangerous species of jellyfish.

Much of our northern coastlines are remote and wild, to say the least. Being in a remote location simply adds to the dangers from animals such as the **Salties**, **Bull Sharks** and venomous jellyfish. However, preventative measures are increasing, from net-enclosed beaches, better signage, jugs of vinegar placed along swimming beaches (for rapid first aid) to good old-fashion awareness - helped I hope by the book you are reading right now.

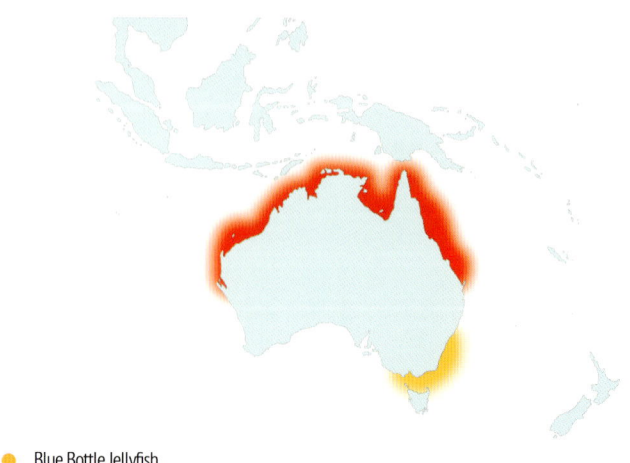

- Blue Bottle Jellyfish
- Box Jellyfish / Irukandji Box Jellfish

THIS PAGE
Floating on the surface are two distinctive Australian Blue Bottles, sailing the high seas. Below the surface are two larger transparent Box Jellyfish and three thumb-nail sized Irukandji Jellyfish.

Blue-ringed Octopus

As an avid snorkeler, I've encountered many Blue-ringed Octopuses in the wild. Inhabiting rockpools and shallow reefs in Port Philip Bay, I feel so privileged to have these octopuses on my doorstep to film and observe their behaviour. They're pretty small and often hard to spot, but they do let you know when they're annoyed with you. Here's another interesting fact I observed while spearfishing on the Yorke Peninsula. Gutting my fish, I discovered chopped up octopus in the stomach of my catch: so Leather Jackets snack on Blue-Ringed Octopuses. You learn something new every day!

A lot of what we describe as dangerous isn't really too deadly. Most of what I talk and write about is really trying to warn people of the consequences of our actions on wildlife, rather than vice versa.

Take the **Blue-ringed Octopus**. On one hand it's highly venomous: one bite can shut down the muscles that keep you breathing and cause your death within 30 minutes. But on the other hand, violent human encounters with these small aquatic 'death wishes' are uncommon. Deaths from a **Blue-ringed Octopus** bite are again extremely rare. Only 3 deaths have been recorded and, while many other people have been bitten, they have all survived.

There are four highly venomous species of **Blue-ringed Octopus**, found in tide pools and coral reefs from Japan to Australia and scattered through the Pacific and Indian oceans. They eat small crustaceans - including crabs, hermit crabs and shrimp - other small sea animals and fish if they can catch them.

If threatened, a **Blue-ringed Octopus** will flee to a dark ledge or try to camouflage itself. If provoked further, the octopus will display its blue rings. If it is cornered, touched or handled by a person, they are likely to be bitten.

Recently there has been some debate about the **Blue-ringed Octopus's** venom. A bite can result in nausea, respiratory arrest, heart failure, severe and sometimes total paralysis, blindness and can lead to death from suffocation due to paralysis of the diaphragm within minutes if not treated. The debate is about whether their venom is produced by the octopus or by a bacteria living in the salivary glands, forming a symbiotic relationship. The science is not conclusive, so for now it remains the world's most venomous marine creature.

Appearance *They can be identified by their yellowish skin and characteristic blue and black rings which change colour dramatically when the animal is threatened.* **Size** *They are relatively small, growing to between 12 and 20cm across.*

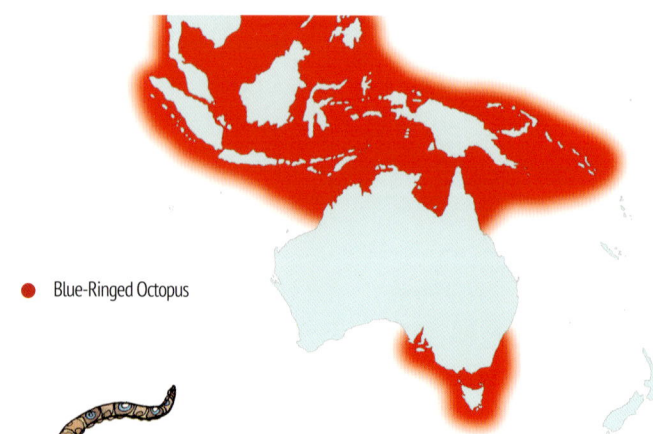

● Blue-Ringed Octopus

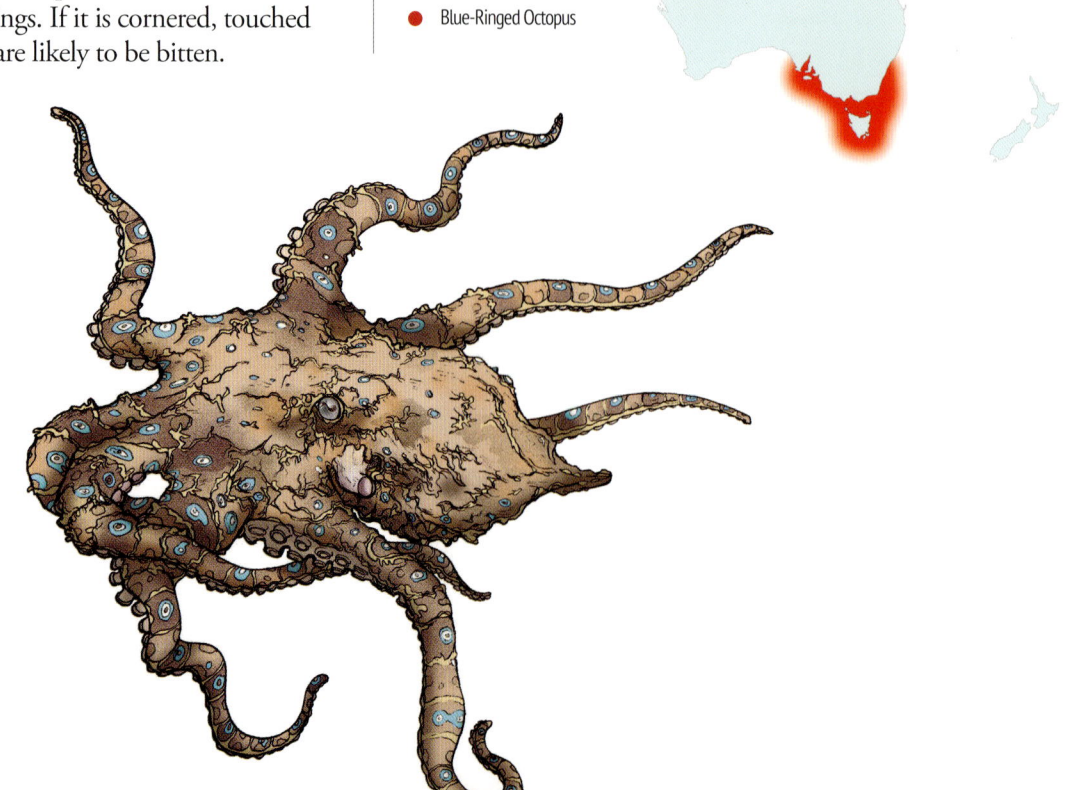

OPPOSITE PAGE
Mostly nocturnal, the Blue-ringed Octopus hunts its prey by the cover of night. However it can be spotted in weedy reefs moving about during the day.

THIS PAGE
Blue-ringed Octopus.

Australia's Lace Monitor – the Goanna

I love Lace Monitors, one of many species of Goanna. I've shared many bush campsites with these prehistoric-looking creatures. They have raided my esky, savaged my bins looking for scraps and ripped through my tent foraging for food. Because they're diurnal (active during the day), I've seen big males fighting, climbing trees to hunt for baby possums and raiding bird nests. I've also witnessed these scavenging carrion feeders eating from a whale carcass washed up on a Gippsland beach. They truly are the 'garbage guts' of the bush.

There are 28 known species of **Goanna** found throughout the Australian mainland (with the exception of Tasmania). They vary in size, but the largest goannas are the **Heath Goanna**, **Sand Goanna**, **Perentie** and **Lace Monitor**, found mainly in the south-east of the country.

Goannas have been around for millions of years. They have a totemic place in our Indigenous people's culture including in art, ceremony, history, dreamtime stories and, once, as an important food source. The word 'goanna' actually originates from a morphing of the word 'iguana' by the English settlers. These carnivorous reptiles inhabit a wide variety of bushland, deserts and coastal forests. Excellent tree climbers, excavators and swimmers, they are both predator and scavenger. They play an important role in our ecosystems, controlling the populations of several species and removing carcasses, stopping the spread of disease.

Lace Monitors often lounge about on lofty limbs in large trees, sunning themselves, safe from dogs and humans. Using the sun to ferment, liquify and digest their food, they heat and energise their primordial bodies like mobile compost bins. Long forked tongues flicker as they taste the air, sensing predators, rivals, potential mates and their next dinner.

With formidable claws, strong jaws and a raspy bite it's easy to see why these guys are dangerous: they can certainly do some damage if handled or threatened. Interestingly though, before 2005, lingering infections from a **Goanna's** bite were thought to be a consequence of their saliva being rife with bacteria from eating dead things. It's still being debated, but it is now thought that some **Goannas** have venom glands similar to snakes.

Yes **Goannas** are venomous! Not enough venom to cause serious harm and, without fangs, they can't inject their venom into your skin like snakes can. In fact, they rarely attack humans, unless provoked. A large **Goanna** will grow to around 2 metres long, and can reportedly fracture a human forearm, as well as cause severe bleeding and infection, which definitely includes them on our dangerous list.

● Lace Monitor

THIS SPREAD
The large lumbering Lace Monitor. At nearly 2m long, this young male Goanna has a striking lace-pattern in his thick rough skin to camouflage his presence in the dense dry scrub.

Bardick Snake

The shy, reclusive Bardick Snake is only found in three places, the semi-arid landscapes of the Eyre Peninsula, the Mallee regions (including eastern South Australia and north-western Victoria) and south-western NSW.

You may not have heard of the **Bardick Snake**. A short, distinctively stout snake with non-glossy olive-grey to rich reddish brown scales. It's quite venomous, in fact considered as venomous as the **Death Adder**. Although it's usually not considered too dangerous to humans, due to its shy nature, it can become very defensive if disturbed. **Size** *Averages about 40cm long, although it can reach 70cm.*

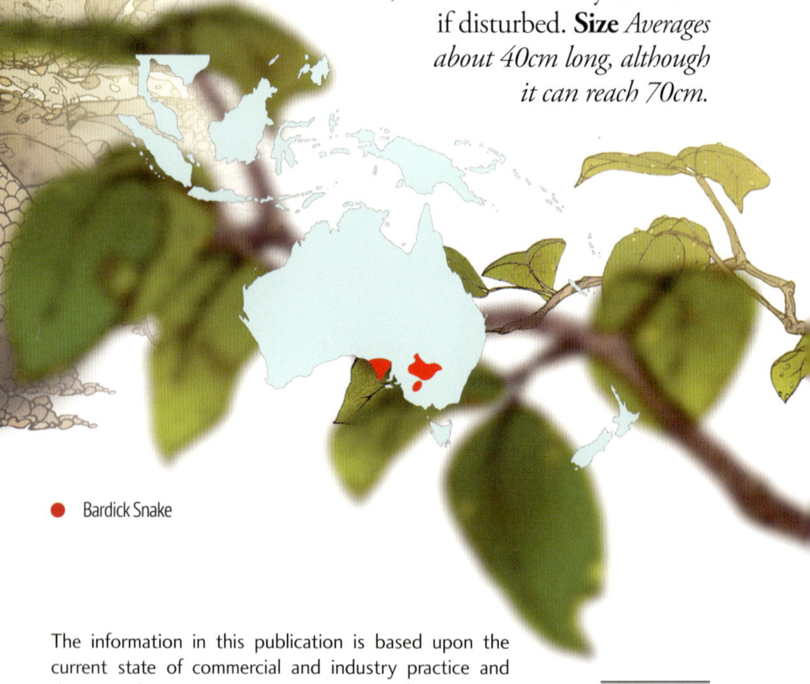

• Bardick Snake

Woodslane Press Pty Ltd
10 Apollo Street
Warriewood, NSW 2102
Email: info@woodslane.com.au
Tel: 02 8445 2300
Website: www.woodslanepress.com.au

Published in Australia in 2022 by Woodslane Press

Text copyright © 2022 Myke Mollard
Illustrations copyright © 2022 Myke Mollard
The moral rights of the author and illustrator have been asserted.

This work is copyright. All rights reserved. Apart from any fair dealing for the purposes of study, research or review, as permitted under Australian copyright law, no part of this publication may be reproduced, distributed, or transmitted in any other form or by any means, including photocopying, recording, or other electronic or mechanical methods, without the prior written permission of the publisher. For permission requests, write to the publisher, addressed "Attention: Permissions Coordinator", at the address above.

Printed in China by Hang Tai
Designed by Myke Mollard
Special thanks to Dennis Jones, James Beckingham and Yiwen Yao for all the enthusiasm and support in making this dream happen.

The information in this publication is based upon the current state of commercial and industry practice and the general circumstances as at the date of publication. Every effort has been made to obtain permissions relating to information reproduced in this publication. The publisher makes no representations as to the accuracy, reliability or completeness of the information contained in this publication. To the extent permitted by law, the publisher excludes all conditions, warranties and other obligations in relation to the supply of this publication and otherwise limits its liability to the recommended retail price. In no circumstances will the publisher be liable to any third party for any consequential loss or damage suffered by any person resulting in any way from the use or reliance on this publication or any part of it. Any opinions and advice contained in the publication are offered solely in pursuance of the author's and publisher's intention to provide information, and have not been specifically sought.

 A catalogue record for this book is available from the National Library of Australia

 **MIX** Paper from responsible sources FSC® C023121

THIS PAGE
The delicate Bardick Snake nestled in the underbrush.

North

Australia
Including Surrounding Islands & Territories

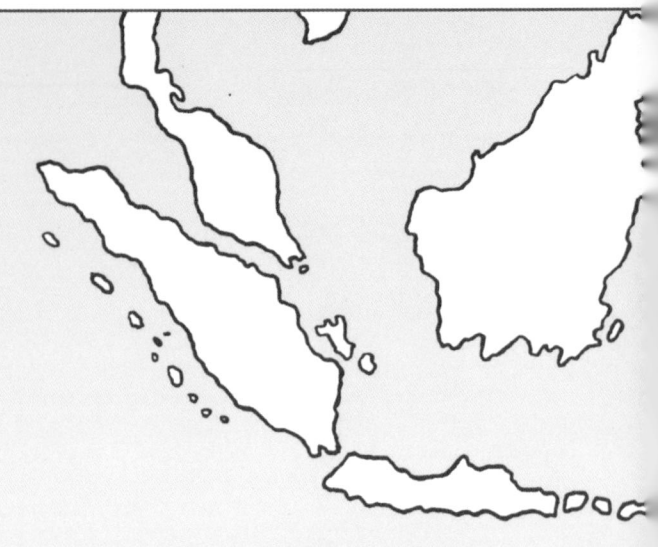

- Christmas Island

Cocos Keeling Islands

Exmouth Gulf
Ningaloo Reef

Shark Bay

Margaret River

Indian Ocean

- Heard Island

Antarctica